# Figure & Form

## VOLUME II

### An Anatomical Coloring Book For Artists
### with Photographic Glossary

# Figure & Form

## VOLUME II

## *An Anatomical Coloring Book For Artists with Photographic Glossary*

## Lu Bro

WCB Brown & Benchmark

**Book Team**

Production Editor *Kay J. Brimeyer*
Designer *Elise A. Burckhardt*
Photo Editor *Sharon A. LaPrell*
Art Editor *Kenneth E. Ley*
Visuals/Design Developmental Consultant *Marilyn A. Phelps*

**WCB Brown & Benchmark**
A Division of Wm. C. Brown Communications, Inc.

**Wm. C. Brown Communications, Inc.**

**Front Cover:** Leonardo da Vinci (1452–1519). Proportional study of man in the manner of Vitruvius. Circa 1487. Pen and ink. Venice Accademia.

**Back Cover:** Leonardo da Vinci (1452–1519) "Nude Man Study, back to a Spectator." Red Chalk, $10\frac{5}{8} \times 6\frac{5}{16}$. Windsor Castle, Royal Library. © Her Majesty Queen Elizabeth II.

# PREFACE

▼

The second volume of *Figure and Form* is intended to support the practices encouraged in the first volume.

The opening section of this volume, "An Anatomical Coloring Book for Artists," is a nuts-and-bolts sequential grouping of anatomical pictures that takes you through the human body, part by part, head to toe, surface to bone. The information, layout, and sequencing, along with your participation, should give you a convenient reference guide to making sense of muscles.

We begin with an abbreviated history of life drawing. We move to a series of plates, color coding as we go through the *axial* skeleton (pertaining to or forming an *axis*) and the corresponding muscles with yellow felt-tip highlighters, and color coding the *appendicular* skeleton (pertaining to the *appendages*) and corresponding muscles with blue highlighters. Those bones that overlap the axial and appendicular skeleton and corresponding muscles are color coded with green highlighters.

The first freehand drawings are initiated here. You should have an articulated skeleton along with a live model available in your studio or classroom. The most productive process I have found by which students learn about the human body is through drawing the blind contour on one sheet of paper, drawing the gesture/sighting/volumes on another sheet, and imposing bones into the flesh in yet another drawing.

Beginning from the head and progressing through the feet, each body unit is laid out on facing pages with names, actions, origins, and insertions. You color code your own way, defining any color system you choose. However, a *consistent* color coding for muscles, origins, and insertions can help you retain information most effectively.

Again, at the end of each body sequence are the three basic freehand drawing practices, the contour, the gesture/sighting/volumes, and the imposition of bones into flesh, for use with both the articulated skeleton and a live model.

A closer look at facial muscles and features follows.

The long and thorough sequence of body muscles in motion should be used alongside a live model in your studio or classroom. The model should duplicate each pose in the pictures so that you can begin with the gesture/sighting/volumes of the model, then, with a tracing overlay, you can draw in his or her muscles using the anatomical pictures as a guide.

The first part concludes with a series of anatomical drawings by master artists. The pictures are selected so that you might see both male and female. You can study various types and kinds of features, bodies and limbs, and you can see as many sides of those units as is possible within a drawing book of this scope.

The second part of *Figure and Form*, Volume II, "People and Parts," is a visual glossary, a series of photographs of actual models, male and female, posed similarly but alternating left side to right side. The glossary tries to move past the limitation of having only right or left body positions, forcing the artist to transpose.

The aggravation of not having a model to pose for the urgent right hand or the left foot you need when you are working on a drawing or painting at midnight in your studio (wherever that is) can bring one to grief. My husband tried to accommodate me one time.

I badly needed a highly foreshortened left thigh, leg, and foot of a woman kneeling on a five-inch elevation, like a rock, four feet below eye level. I had looked through my own rather extensive figure drawing library. Nothing worked. So, my husband grudgingly agreed to change into shorts and to pose. I knew I would have to adjust the musculature from male to female later. I placed him with one knee on a short stool while he placed his other leg straight out to his side as a stiffened support in this awkward tripod stance. Busy with posing him and finding my drawing position about six feet in front and a little to the side of him, I had not realized he himself was comfortably making do by eating a bowl of ice cream while watching a football game. The drawing/painting was accepted into three national shows.

Husbands and wives, boyfriends and girlfriends, are not always around when you need a model. Maybe some of this volume's photographs will help you when you are caught without reference to real limb or torso.      ▲

# Figure & Form

## VOLUME II

*An Anatomical Coloring Book For Artists*
*with Photographic Glossary*

# AN ANATOMICAL COLORING
# BOOK FOR ARTISTS

The nude as figure and form can be seen in nearly all the periods of achievement in Western art, simply because the human being is our measure of the universe. Only the modes differ.

From the *Venus of Willendorf*, c. 15,000–10,000 B.C., to Italian Renaissance masterpieces such as Michelangelo's *Sistine Ceiling*, c. 1508–1512, to Jasper Johns' actual hand and face transfers in his *Study for Skin I*, 1962, artistic concerns have revolved around the figure, often the nude figure. During the Middle Ages, art students were apprenticed to masters in whose workshops they practiced. In the middle of the 1400s students began working from both antique statues and live models. By the 1500s artistic inquiry became an intellectual discipline and not just a manual craft. The first established academy was founded in Rome in 1593.

Much groundwork had been laid before, however, by dedicated artists who dug up, dissected, and drew body parts. Their need to know overrode the legal ramifications had they been caught.

Antonio del Pollaiuolo (c. 1432–98) was one of the earliest to realize that body movement demanded detailed anatomical information. H. W. Janson suggests Pollaiuolo might have been the first to "dissect human bodies, a practice then uncommon even in medical schools."[1] *The Battle of Ten Naked Men*, an engraving, c. 1465–70, includes ten male figures in motion with dramatic, almost flayed or exposed, musculature. Each man's face is individual and expressive as well.

Since 1593, academies or art schools have dwelled extensively on the figure. Where anatomy should fit for art students is in an informed and expressive stroke. An informed stroke carries information, which, for figure drawing, means knowing about bones, muscles, and joints and what they can and cannot do. Anatomical information is not ipso facto esthetic. Anatomical drawings can be as boring as almost any display of skill alone. But skills are useful and often necessary for growth in expression. Through an understanding of anatomy and its functions, a student can eventually ease into a stylistic range from realism (detail) to abstractionism (succinctness) for this reason: the stroke will carry both informed knowledge and informed feeling. An artist needs knowledge *and* feeling to shape evocative images. To understand the figure and the form by which it is presented is to communicate an affective statement. That statement or meaning in a work is what moves us as viewers, what commands our thought, and finally what determines the capacity of a work to be considered art.

From the ensuing presentation and its use of color coding, the student should expect to master anatomical information. We begin with bones, separating the skeleton into two units, the axial (what rotates on an axis) and the appendicular (the appendages). Muscles pertaining to those skeletal units are color coded with yellow and blue. Those muscles that overlap both skeletal units are colored green.

The anatomy section is separated into subsections, the head, the torso, the arm, and so on. There are four pictures for each subsection. A surface anatomy is one. A similar image flayed with musculature is the second. A coloring book for your own color coding is the third. And the fourth is a skeleton, again color coded, but for origins and insertions of muscles. After each subsection three very basic types of drawing exercises on separate drawing papers are suggested. These exercises should be done while looking at an attenuated skeleton and a model. The first exercise is working with blind contour. The second structures the figure through gesture, sighting, and volumes. And the third superimposes the bones into the flesh of the model.

A discussion about facial expressions includes using tracing overlays that help identify muscles of the face.

Movements of the body show the model in a number of positions. Each of the positions can be duplicated by a model in the classroom. If you begin with the gestural structure and volumes, placing muscles using a tracing overlay should be more familiar for you by then.

Finally, observing what some of the great artists have done with anatomical information is important because they combined skill with expression in many of the illustrations. In some you can see how they too, were struggling for accuracy, so that when they needed to apply skill, it became the catapult to content and form, or figure and form.

An extra word here about coloring the anatomical plates. Coloring can be used as an aid to memorizing parts of the body. By coding muscles and their origins and insertions to bones with colors you can choose, your learning curve should be broader than by a rote memory approach. A helpful hint: once you choose the color for one muscle, try to retain that color for that muscle throughout the exercises. The human body has many, many muscles, far more than there are bones. Some plates may require as many as 17 different colors. As much as possible, you should retain the same color for the same muscle throughout. A large box of assorted colored pencils will help you retain distinctive color coding within this section.  ▲

1. H. W. Janson, *History of Art*, Prentice Hall, Inc. and Harry N. Abrams, Inc., (July, 1969): 340.

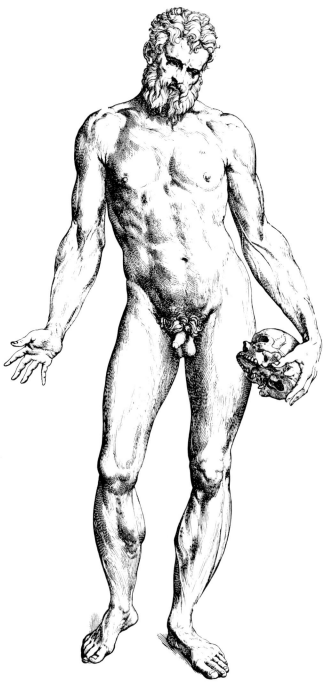

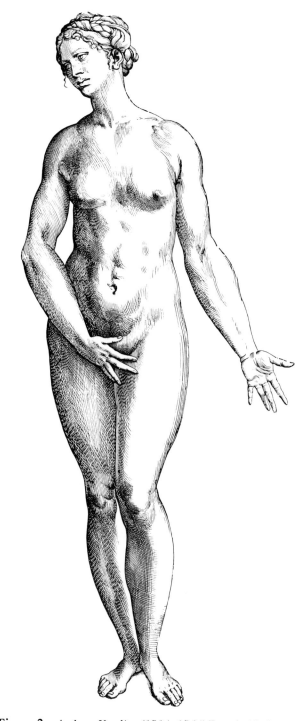

***Figure 1*** *Andreas Vesalius. (1514–1564) Male Nude. As appeared in* The Illustrations from the Works of Andreas Vesalius of Brussels, *by J. B. de C. M. Saunders and Charles D. O'Mally. Dover Publications, Inc., New York.*

***Figure 2*** *Andreas Vesalius (1514–1564) Female Nude. As appeared in* The Illustrations from the Works of Andreas Vesalius of Brussels, *by J. B. de C. M. Saunders and Charles D. O'Mally. Dover Publications, Inc., New York.*

***Figure 3*** *Antonio del Pollaiuolo.* Battle of Ten Naked Men. *Engraving, c. 1465–70. The Metropolitan Museum of Art, NY. [Joseph Pulitzer Bequest, 1917.]*

# THE SKELETON

The skeleton has two major divisions, the AXIAL SKEL-ETON (pertaining to or forming an axis) and the APPEN-DICULAR SKELETON (pertaining to appendages).

## UPPER BODY

### Axial

Braincase—Skull

Backbone—Vertebrae

Breastbone—Sternum

Ribs—Costals

### Appendicular

The shoulder unit—Pectoral Girdle

Collarbone—Clavicle

Shoulder blades—Scapulae

Arms—Humerus

Forearm—Radius & Ulna

Wrists—Carpals

Hands—Metacarpals and Phalanges

Hip and sit bones—Pelvic Girdle
     Ilium
     Ischium
     Pubis

## LOWER BODY

Thigh—Femur

Leg—Tibia and Fibula

Ankles—Tarsals

Feet—Metatarsals and Phalanges

On the next three *schematic line drawings:*

Color the AXIAL SKELETON *YELLOW* AND underline/highlight the name of the bones on the list that correspond to the Axial Skeleton.

Color the APPENDICULAR SKELETON *BLUE* AND underline/highlight the names of the bones on the list that correspond to the Appendicular Skeleton.

Later, working with musculature you will see by using similar yellow and blue coordinates for muscles, that some muscles overlap from the axial to the appendicular skeleton. Yellow and blue, mixed, make green. Those overlapping muscles will be coded *GREEN*.

## FRONT VIEW: SKELETON

Those bones associated with the *AXIAL* skeleton: Color them YELLOW.

1 through 6, 11, 20 through 31, 35

Those bones associated with the *APPENDICULAR* skeleton: Color all the remaining bones BLUE.

## SIDE VIEW: SKELETON

Those bones associated with the *AXIAL* skeleton: Color them YELLOW.

1 through 4, 14 through 17, 31 through 36, 38, 42, 43

Those bones associated with the *APPENDICULAR* skeleton: Color all the remaining bones BLUE.

## BACK VIEW: SKELETON

Those bones associated with the *AXIAL* skeleton: Color them YELLOW.

1 through 4, 15 through 18, 21, 23, 26, 27, 29

Those bones associated with the *APPENDICULAR* skeleton: Color all the remaining bones BLUE.

On the next three schematic line drawings:

Color the muscles *associated* with the *AXIAL* skeleton YELLOW and underline/highlight those muscles on the corresponding list.

Color the muscles *associated* with the *APPENDICULAR* skeleton light BLUE and underline/highlight those muscles on the corresponding list.

Color the muscles *associated* with *BOTH AXIAL AND APPENDICULAR,* GREEN.

## FRONT VIEW: MUSCLES

Those muscles associated with: the *AXIAL* skeleton: Color them YELLOW.

1, 2, 3, 22, 23, 24

Those muscles associated with: the *APPENDICULAR* skeleton: Color them BLUE.

4 through 21, 28 through 34, 39 through 46, 47 through 68

Those muscles associated with both the *AXIAL* and *APPENDICULAR* skeleton: Color them GREEN.

25 through 27, 35 through 38

## SIDE VIEW: MUSCLES

Those muscles associated with the *AXIAL* skeleton: Color them YELLOW.

1 through 4, 44, 46, 48, 50

Those muscles associated with the *APPENDICULAR* skeleton: Color them BLUE.

6 through 22, 26 through 43, 51 through 53, 55 through 60, 62 through 68

Those muscles associated with both the *AXIAL* and *APPENDICULAR* skeleton: Color them GREEN.

5, 23, 24, 25, 45, 47, 49, 54, 61

## BACK VIEW: MUSCLES

Those muscles associated with the *AXIAL* skeleton: Color them YELLOW.

1, 40, 41

Those muscles associated with the *APPENDICULAR* skeleton: Color them BLUE.

3 through 19, 21, 22, 24 through 39, 44 through 46, 49 through 58

Those muscles associated with both the *AXIAL* and *APPENDICULAR* skeleton: Color them GREEN.

2, 20, 23, 42, 43, 47, 48 ▲

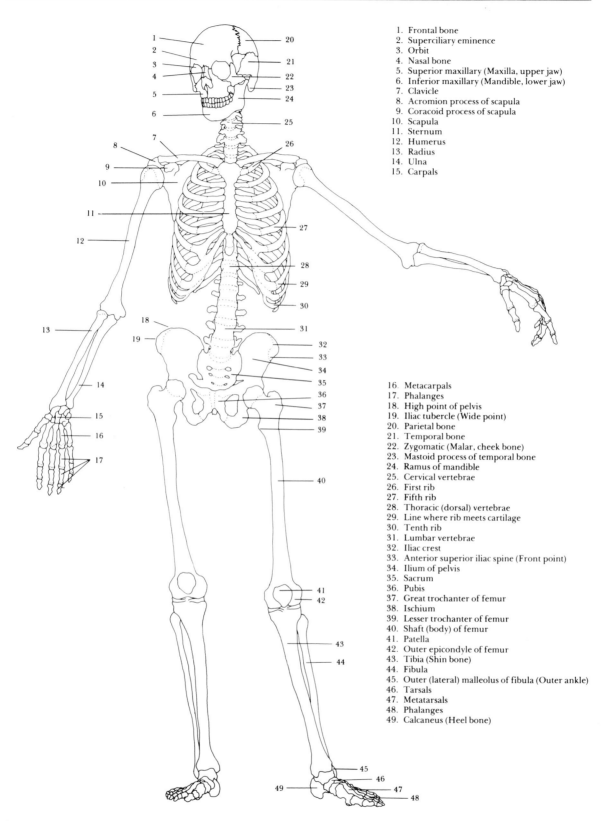

1. Frontal bone
2. Superciliary eminence
3. Orbit
4. Nasal bone
5. Superior maxillary (Maxilla, upper jaw)
6. Inferior maxillary (Mandible, lower jaw)
7. Clavicle
8. Acromion process of scapula
9. Coracoid process of scapula
10. Scapula
11. Sternum
12. Humerus
13. Radius
14. Ulna
15. Carpals

16. Metacarpals
17. Phalanges
18. High point of pelvis
19. Iliac tubercle (Wide point)
20. Parietal bone
21. Temporal bone
22. Zygomatic (Malar, cheek bone)
23. Mastoid process of temporal bone
24. Ramus of mandible
25. Cervical vertebrae
26. First rib
27. Fifth rib
28. Thoracic (dorsal) vertebrae
29. Line where rib meets cartilage
30. Tenth rib
31. Lumbar vertebrae
32. Iliac crest
33. Anterior superior iliac spine (Front point)
34. Ilium of pelvis
35. Sacrum
36. Pubis
37. Great trochanter of femur
38. Ischium
39. Lesser trochanter of femur
40. Shaft (body) of femur
41. Patella
42. Outer epicondyle of femur
43. Tibia (Shin bone)
44. Fibula
45. Outer (lateral) malleolus of fibula (Outer ankle)
46. Tarsals
47. Metatarsals
48. Phalanges
49. Calcaneus (Heel bone)

***Figure 4*** *Albinus (1697–1770). The skeleton, anterior (front). [Albinus on Anatomy. By Robert Beverly Hale and Terrance Coyle. Watson-Guptill Pub., NY. 1979. p. 28.] [Original Source: New York Academy of Medicine. Tables of the Human Body and Tables of the Human Bones.]*

***Figure 5*** *Albinus (1697–1770). The skeleton, anterior (front).* [Albinus on Anatomy. By Robert Beverly Hale and Terrance Coyle. Watson-Guptill Pub., NY. 1979. p. 29.]

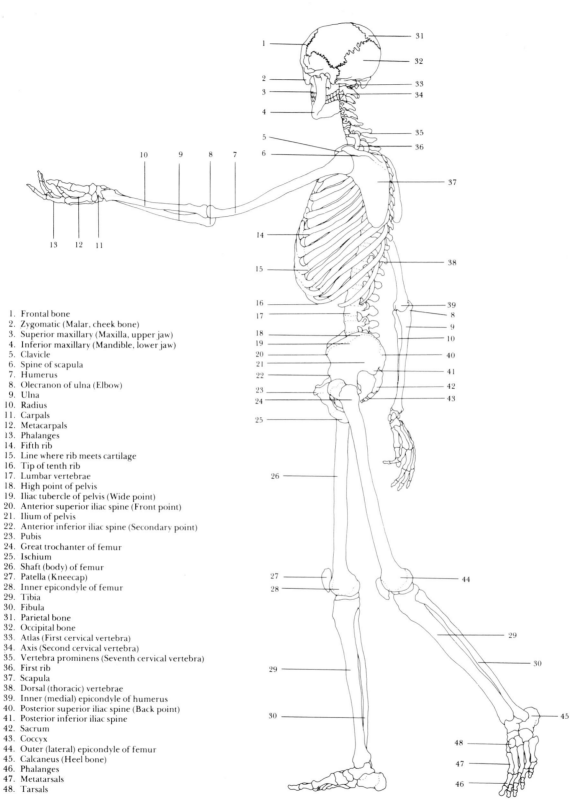

1. Frontal bone
2. Zygomatic (Malar, cheek bone)
3. Superior maxillary (Maxilla, upper jaw)
4. Inferior maxillary (Mandible, lower jaw)
5. Clavicle
6. Spine of scapula
7. Humerus
8. Olecranon of ulna (Elbow)
9. Ulna
10. Radius
11. Carpals
12. Metacarpals
13. Phalanges
14. Fifth rib
15. Line where rib meets cartilage
16. Tip of tenth rib
17. Lumbar vertebrae
18. High point of pelvis
19. Iliac tubercle of pelvis (Wide point)
20. Anterior superior iliac spine (Front point)
21. Ilium of pelvis
22. Anterior inferior iliac spine (Secondary point)
23. Pubis
24. Great trochanter of femur
25. Ischium
26. Shaft (body) of femur
27. Patella (Kneecap)
28. Inner epicondyle of femur
29. Tibia
30. Fibula
31. Parietal bone
32. Occipital bone
33. Atlas (First cervical vertebra)
34. Axis (Second cervical vertebra)
35. Vertebra prominens (Seventh cervical vertebra)
36. First rib
37. Scapula
38. Dorsal (thoracic) vertebrae
39. Inner (medial) epicondyle of humerus
40. Posterior superior iliac spine (Back point)
41. Posterior inferior iliac spine
42. Sacrum
43. Coccyx
44. Outer (lateral) epicondyle of femur
45. Calcaneus (Heel bone)
46. Phalanges
47. Metatarsals
48. Tarsals

**Figure 6** *Albinus (1697–1770). The skeleton, lateral (side).* [Albinus on Anatomy. *By Robert Beverly Hale and Terrance Coyle. Watson-Guptill Pub., NY. 1979. p. 32.*]

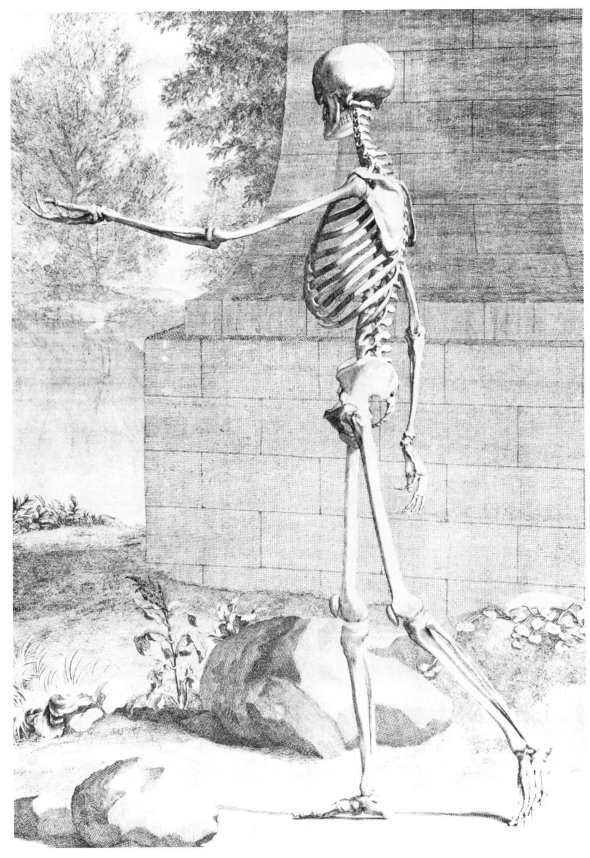

***Figure 7*** *Albinus (1697–1770). The skeleton, lateral (side). [*Albinus on Anatomy. *By Robert Beverly Hale and Terrance Coyle. Watson-Guptill Pub., NY. 1979. p. 33.]*

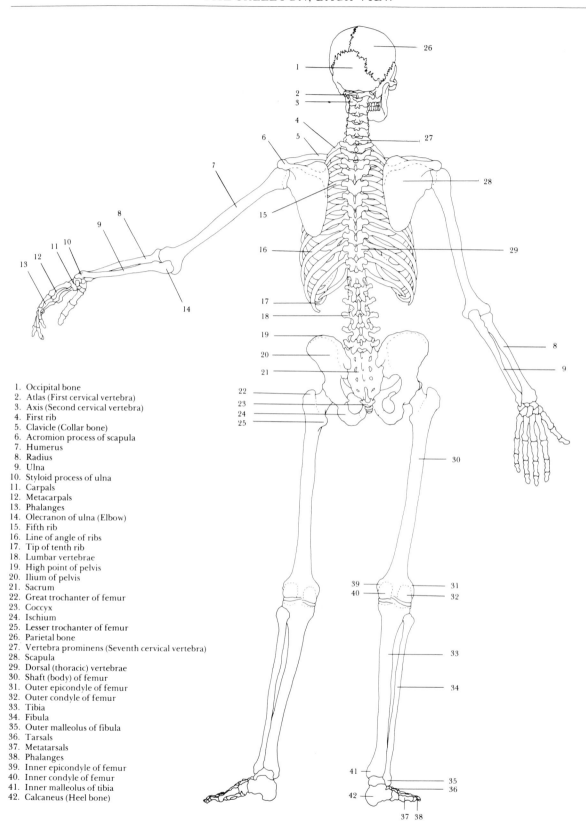

1. Occipital bone
2. Atlas (First cervical vertebra)
3. Axis (Second cervical vertebra)
4. First rib
5. Clavicle (Collar bone)
6. Acromion process of scapula
7. Humerus
8. Radius
9. Ulna
10. Styloid process of ulna
11. Carpals
12. Metacarpals
13. Phalanges
14. Olecranon of ulna (Elbow)
15. Fifth rib
16. Line of angle of ribs
17. Tip of tenth rib
18. Lumbar vertebrae
19. High point of pelvis
20. Ilium of pelvis
21. Sacrum
22. Great trochanter of femur
23. Coccyx
24. Ischium
25. Lesser trochanter of femur
26. Parietal bone
27. Vertebra prominens (Seventh cervical vertebra)
28. Scapula
29. Dorsal (thoracic) vertebrae
30. Shaft (body) of femur
31. Outer epicondyle of femur
32. Outer condyle of femur
33. Tibia
34. Fibula
35. Outer malleolus of fibula
36. Tarsals
37. Metatarsals
38. Phalanges
39. Inner epicondyle of femur
40. Inner condyle of femur
41. Inner malleolus of tibia
42. Calcaneus (Heel bone)

***Figure 8*** *Albinus (1697–1770). The skeleton, posterior (back).* [Albinus on Anatomy. *By Robert Beverly Hale and Terrance Coyle. Watson-Guptill Pub., NY. 1979. p. 30.*]

***Figure 9*** *Albinus (1697–1770). The skeleton, posterior (back). [Albinus on Anatomy. By Robert Beverly Hale and Terrance Coyle. Watson-Guptill Pub., NY. 1979. p. 31.]*

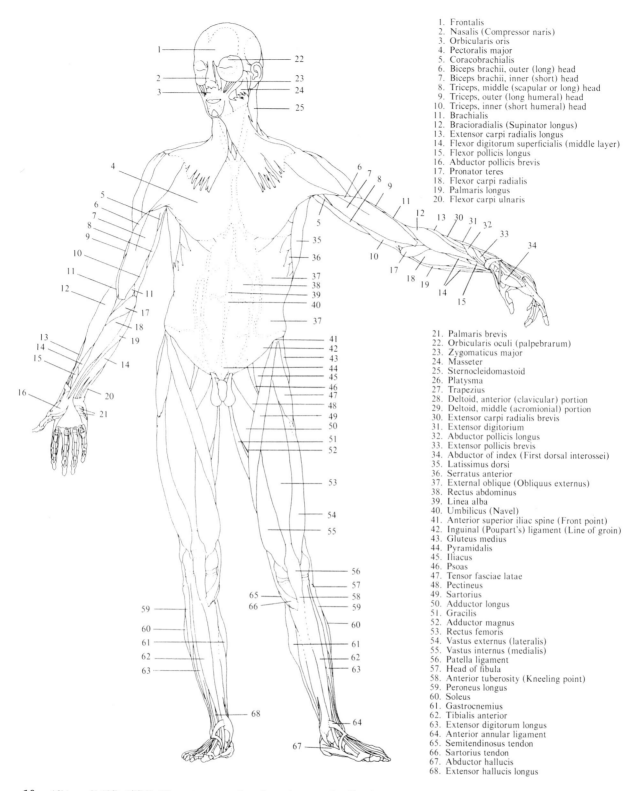

1. Frontalis
2. Nasalis (Compressor naris)
3. Orbicularis oris
4. Pectoralis major
5. Coracobrachialis
6. Biceps brachii, outer (long) head
7. Biceps brachii, inner (short) head
8. Triceps, middle (scapular or long) head
9. Triceps, outer (long humeral) head
10. Triceps, inner (short humeral) head
11. Brachialis
12. Bracioradialis (Supinator longus)
13. Extensor carpi radialis longus
14. Flexor digitorum superficialis (middle layer)
15. Flexor pollicis longus
16. Abductor pollicis brevis
17. Pronator teres
18. Flexor carpi radialis
19. Palmaris longus
20. Flexor carpi ulnaris

21. Palmaris brevis
22. Orbicularis oculi (palpebrarum)
23. Zygomaticus major
24. Masseter
25. Sternocleidomastoid
26. Platysma
27. Trapezius
28. Deltoid, anterior (clavicular) portion
29. Deltoid, middle (acromionial) portion
30. Extensor carpi radialis brevis
31. Extensor digitorium
32. Abductor pollicis longus
33. Extensor pollicis brevis
34. Abductor of index (First dorsal interossei)
35. Latissimus dorsi
36. Serratus anterior
37. External oblique (Obliquus externus)
38. Rectus abdominus
39. Linea alba
40. Umbilicus (Navel)
41. Anterior superior iliac spine (Front point)
42. Inguinal (Poupart's) ligament (Line of groin)
43. Gluteus medius
44. Pyramidalis
45. Iliacus
46. Psoas
47. Tensor fasciae latae
48. Pectineus
49. Sartorius
50. Adductor longus
51. Gracilis
52. Adductor magnus
53. Rectus femoris
54. Vastus externus (lateralis)
55. Vastus internus (medialis)
56. Patella ligament
57. Head of fibula
58. Anterior tuberosity (Kneeling point)
59. Peroneus longus
60. Soleus
61. Gastrocnemius
62. Tibialis anterior
63. Extensor digitorum longus
64. Anterior annular ligament
65. Semitendinosus tendon
66. Sartorius tendon
67. Abductor hallucis
68. Extensor hallucis longus

**Figure 10** *Albinus (1697–1770). The outermost order of muscles, anterior (front).* [Albinus on Anatomy. *By Robert Beverly Hale and Terrance Coyle. Watson-Guptill Pub., NY. 1979. p. 34.]*

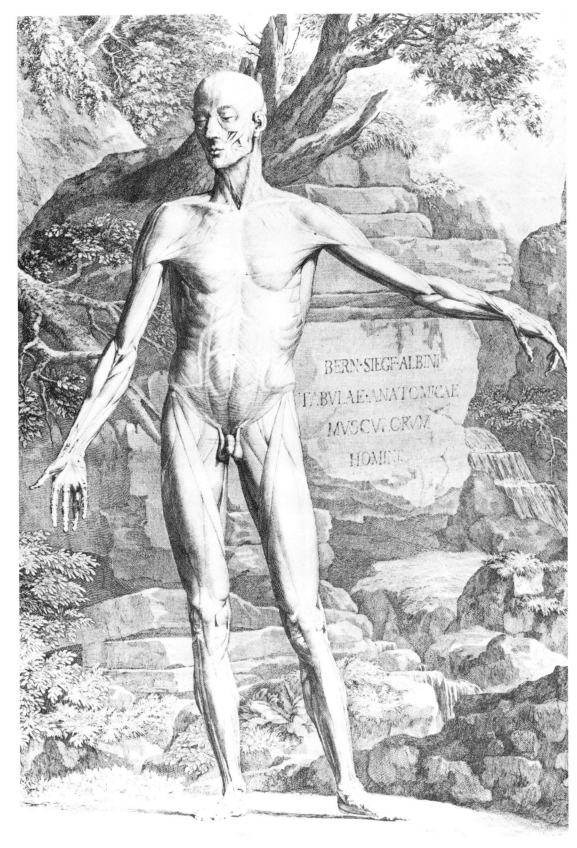

**Figure 11**   *Albinus (1697–1770). The outermost order of muscles, anterior (front).* [Albinus on Anatomy. By Robert Beverly Hale and Terrance Coyle. Watson-Guptill Pub., NY. 1979. p. 35.]

1. Frontalis
2. Auricularis
3. Temporalis
4. Orbicularis oculi (palpebrarum)
5. Platysma
6. Deltoid, middle (acromionial) portion
7. Deltoid, posterior (scapular) portion
8. Biceps brachii, outer (long) head
9. Brachialis
10. Brachioradialis (Supinator longus)
11. Flexor carpi radialis
12. Triceps, middle (scapular or long) head
13. Triceps, outer (long humeral) head

14. Olecranon of ulna (Elbow)
15. Extensor carpi radialis longus
16. Flexor carpi ulnaris
17. Extensor carpi ulnaris
18. Extensor carpi radialis brevis
19. Extensor digitorum
20. Abductor pollicis longus
21. Extensor pollicis brevis
22. Pectoralis major
23. Serratus anterior
24. External oblique (Obliquus externus)
25. Rectus abdominus
26. Anterior superior iliac spine
27. Tensor fasciae latae
28. Adductor longus
29. Sartorius (left leg)
30. Rectus femoris
31. Sartorius (right leg)
32. Vastus internus (medialus)
33. Gracilis
34 Semimembranosus
35. Semitendinosus
36. Patella ligament
37. Gastrocnemius, inner head
38. Soleus
39. Plantaris tendon
40. Tibialis anterior
41. Flexor digitorum longus
42. Flexor hallucis longus
43. Achilles tendon
44. Occipitalis
45. Trapezius
46. Splenius capitis
47. Sternocleidomastoid
48. Splenius cervicis
49. Levator scapulae
50. Vertebra prominens (Seventh cervical vertebra)
51. Teres minor
52. Infraspinatus
53. Teres major
54. Latissimus dorsi
55. Triceps, inner (short humeral) head
56. Biceps brachii, inner (short) head
57. Pronator teres
58. Palmaris longus
59. Flexor digitorum superficialis (middle layer)
60. Gluteus medius
61. Gluteus maximus
62. Biceps femoris, long head
63. Vastus externus
64. Biceps femoris, short head
65. Extensor digitorum longus
66. Gastrocnemius, outer head
67. Peroneus brevis
68. Peroneus tertius

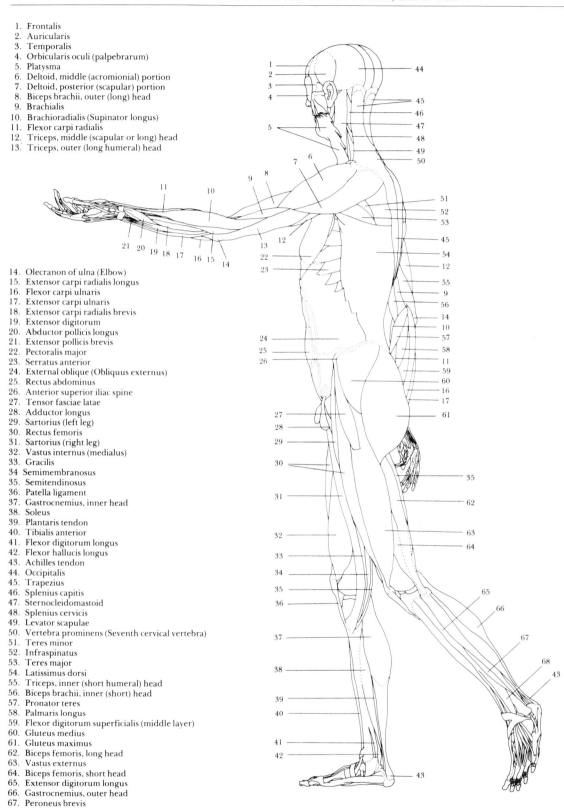

**Figure 12**  *Albinus (1697–1770). The outermost order of muscles, lateral (side).* [Albinus on Anatomy. *By Robert Beverly Hale and Terrance Coyle.* Watson-Guptill Pub., NY. 1979. p. 38.]

***Figure 13*** *Albinus (1697–1770). The outermost order of muscles, lateral (side).* [Albinus on Anatomy. *By Robert Beverly Hale and Terrance Coyle. Watson-Guptill Pub., NY. 1979. p. 39.*]

1. Occipitalis
2. Sternocleidomastoideus
3. Deltoid, posterior (scapular) portion
4. Deltoid, middle (acromionial) portion
5. Triceps, middle (scapular or long) head
6. Triceps, outer (long humeral) head
7. Brachialis
8. Brachioradialis (Supinator longus)
9. Extensor carpi radialis longus
10. Extensor carpi radialis brevis
11. Extensor digitorum
12. Extensor digiti minimi
13. Extensor carpi ulnaris
14. Triceps, inner (short humeral) head
15. Palmaris longus
16. Anconeus

17. Flexor digitorum profundus (deep layer)
18. Flexor carpi ulnaris
19. Flexor digitorum superficialis (middle layer)
20. External oblique
21. Iliac crest
22. Gluteus medius
23. Gluteus maximus
24. Tensor fasciae latae
25. Adductor magnus
26. Vastus externus (lateralis)
27. Biceps femoris, long head
28. Biceps femoris, short head
29. Popliteal space
30. Plantaris
31. Head of fibula
32. Gastrocnemius, outer head
33. Gastrocnemius, inner head
34. Soleus
35. Peroneus longus
36. Peroneus brevis
37. Flexor hallucis longus
38. Superior extensor retinaculum (annular ligament)
39. Abductor digiti minimi
40. Frontalis
41. Orbicularis oculi (palpebrarum)
42. Splenius capitis
43. Trapezius
44. Infraspinatus
45. Teres minor
46. Teres major
47. Rhomboideus major
48. Latissimus dorsi
49. Abductor pollicis longus
50. Extensor pollicis brevis
51. Gracilis
52. Semimembranosus
53. Semitendinosus
54. Sartorius
55. Vastus internus (medialis)
56. Achilles tendon
57. Calcaneus (Heel bone)
58. Tibialis posterior

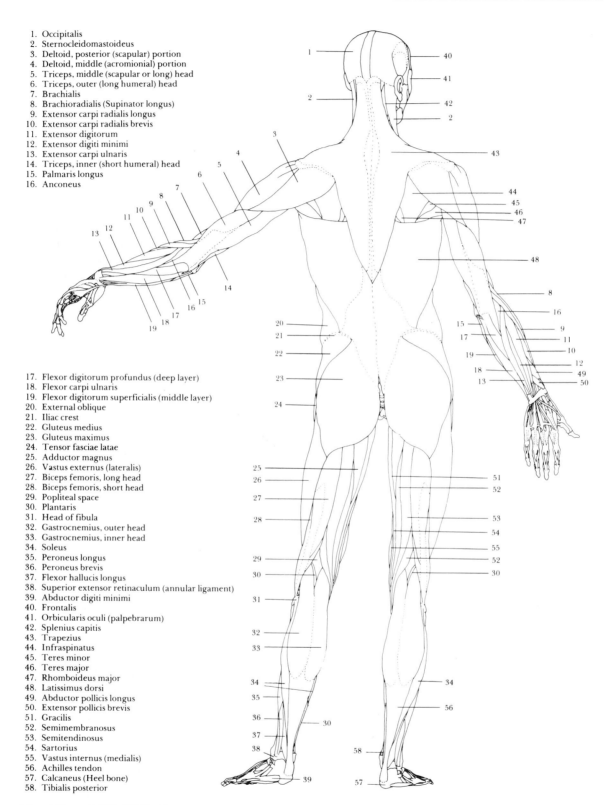

**Figure 14**  *Albinus (1697–1770). The outermost order of muscles, posterior (back).* [Albinus on Anatomy. By Robert Beverly Hale and Terrance Coyle. Watson-Guptill Pub., NY. 1979. p. 36.]

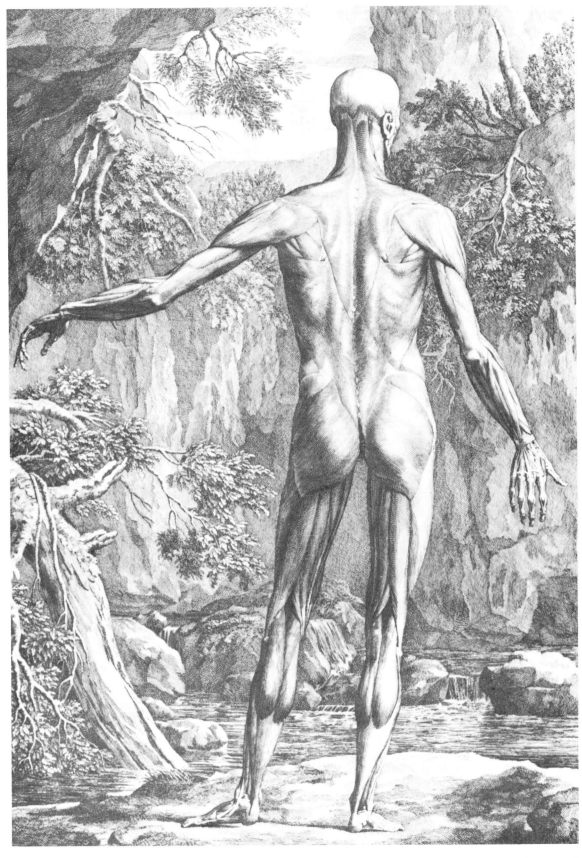

***Figure 15*** *The outermost order of muscles, posterior (back).* [Albinus on Anatomy. *By Robert Beverly Hale and Terrance Coyle. Watson-Guptill Pub.,* NY. 1979. p. 37.]

# THE ANATOMICAL DRAWING STUDIES

Three decades of teaching anatomical skills have forged a few special approaches that can serve students over and over again. For anatomy the clues are SENSITIZING the eye with blind contour, STRUCTURING the figure through gesture/sighting/volumes and SUPERIM-POSING the bones into the flesh.

Because these three approaches are so very rich in informing your eye, the sequencing in the exercises will be consistent throughout this section. Given enough time nearly anyone can draw edges and shade the physical form with a pencil making that form appear to be a bone, a head, a hand. But in a reasonably short time a good student will progress farthest if he/she can:

1. SENSITIZE the eye to line with *blind contour.*
2. STRUCTURE the figure with *gesture/sighting/volumes.*
3. SUPERIMPOSE bones into flesh.

Sensitivity, structure, and superimposition are very important denominators for skill development in drawing. You will guide yourself well through these pages by applying these three exercises at the end of each unit on the body. Sample drawings will show you just what the possibilities are.

*Draw each problem on separate sheets as individual exercises.*

If a student does not understand body parts as geometric units, the tendency to draw two edges using *only* value as a device to suggest volume is very strong. Rubbing or smearing values between edges suggests an uninformed, not very disciplined, and often below average drawing. Good draftsmanship incorporates facility with sighting/proportions, volumes/mass, and sensitive line as well as value.

So far in the anatomy section you have been color coding in your book the axial and appendicular skeletal drawings of Albinus. Now move to a sheet of drawing paper. Use a 2B pencil and an articulated skeleton you might have in your classroom.

Drawing blind means to "draw under cover." Draw on one paper while holding another sheet of paper above that drawing so you cannot see your drawing hand nor the image being drawn. Begin with the first of the three exercises. Draw each bone in the skeleton trying to name to yourself the bones as you go. More importantly, keep your eyes focused ON THE BONES.

The second exercise needs a second sheet of paper and a sharpened 2B pencil and you may resume your place before the skeleton. For these exercises, try to incorporate the whole skeleton on your page. We will get to parts later.

Referring now back and forth from your paper to the skeleton sight the whole skeleton, drawing vertical, horizontal, diagonal or curvilinear lines for the *direction* of the bones. Drawing the *direction* the bones lie presses you to see their length and the general ratios and proportions of

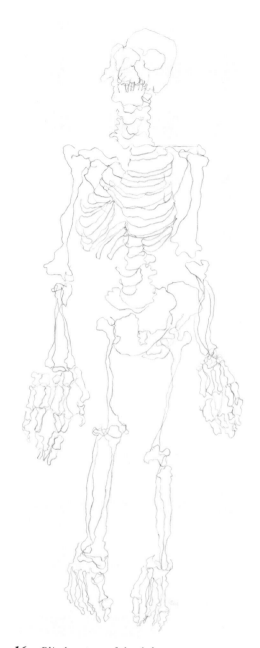

***Figure 16*** *Blind contour of the skeleton.*

the whole group of articulated bones. Remember to do the full form. Once the basic lines of the full skeleton are drawn, go back into parts, sighting the hand, for instance, to something else across/horizontally and up and down/vertically. Begin with the skull, working down (or vertically) and across (or horizontally), constantly sighting one bone to another. Ask yourself, is one scapula higher than the other? Does one shoulder rest lower than the other? Is the sternum at an angle from the clavical, or straight up and down? Where are the elbows relative to the hips, above or below? What is the angle of the hip bones to the backbone? Look vertically and horizontally, correcting as you go. SIGHTING is the issue, not edges, not bones, just

*Figure 17*  *A gesture, sighting, and volumes of the skeleton.*

*Figure 18*  *The superimposition of bones into the body.*

sighting. You should be reducing all the bone shapes to lines—verticals, horizontals, diagonals, or curvilinears, first. Once that is done, go back throughout the skeletal body parts looking for volumes in the bones.

Sometimes the bones will appear flat, sometimes quite rounded. Use cross contour lines and geometric volumes to help infer that the bone, indeed, does go all the way around. The example may help you understand what you need to do. Bear in mind we are not talking about stylistic issues here. Students often are so eager to move ahead with what they perceive as their style that they bypass the discipline of sensitivity, sighting, and superimposing, all of which has very little to do directly with style. These exer-

cises have more to do with practicing the scales of drawing just as one would practice the scales of music on a piano. Repetition and reinforcing what we know helps.

The last of the three exercises is to draw the actual body part from the model, using sighting methods mentioned earlier for accurate ratios, then an external contour line, drawn as you refer to model and paper. The skeleton may well not fit the scale of the model's body and you may need to adjust length or width of bones. But as much as you can, transfer the bones into the drawing of the model's body, reminding yourself of your eye level and the volumes you see relative to that perspective.                    ▲

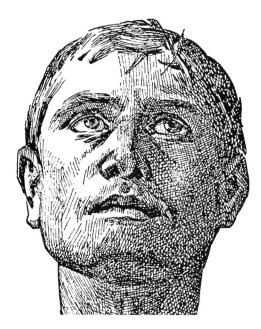

***Figure 19***   *Surface head, anterior (front). As appeared in* An Atlas of Anatomy for Artists. *By Fritz Schider. Dover Publications, Inc. NY.*

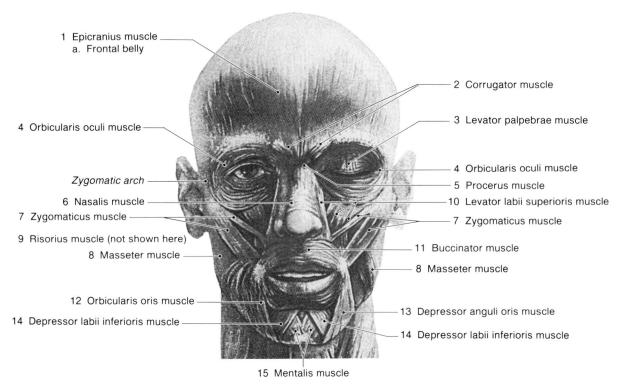

1 Epicranius muscle
   a. Frontal belly

2 Corrugator muscle

3 Levator palpebrae muscle

4 Orbicularis oculi muscle

4 Orbicularis oculi muscle

5 Procerus muscle

*Zygomatic arch*

10 Levator labii superioris muscle

6 Nasalis muscle

7 Zygomaticus muscle

7 Zygomaticus muscle

9 Risorius muscle (not shown here)

11 Buccinator muscle

8 Masseter muscle

8 Masseter muscle

12 Orbicularis oris muscle

13 Depressor anguli oris muscle

14 Depressor labii inferioris muscle

14 Depressor labii inferioris muscle

15 Mentalis muscle

***Figure 20***   *Stephen Rogers Peck. Flayed head, anterior (front). From* An Atlas of Human Anatomy for the Artist *by Stephen Rogers Peck. Copyright © 1951 by Oxford University Press, Inc.; renewed 1979 by Stephen Rogers Peck. Reprinted by permission of Oxford University Press, Inc.*

| Name | Action | Origin | Insertion |
|---|---|---|---|
| 1. EPICRANIUS:<br>(a) Frontal belly<br>(b) Occipital belly | (a) Moves scalp backwards<br>(b) Wrinkles forehead | (a) Sheath of occipital belly<br>(b) Occipital bone and mastoid process | (a) Skin of eyebrows<br>(b) Sheath of frontal belly |
| 2. CORRUGATOR | Draws eyebrows toward midline | Nasal part of frontal bone and tissue above eyebrow | Root of nose |
| 3. LEVATOR PALPEBRAE | Opens upper eyelid | Orbital roof | Skin of upper eyelid |
| 4. ORBICULARIS OCULI | Firmly closes eyelid | Rim of orbit | Edge of lidslit |
| 5. PROCERUS | Wrinkles skin horizontally at bridge of nose | Nasal bones | Skin in between eyebrows |
| 6. NASALIS | Compresses nostrils | Maxilla nasal cartilage | Sheath covering nostrils |
| 7. ZYGOMATICUS | Moves corners of mouth outward and upward | Zygomatic bone | Orbicularis oris and skin at corners of mouth |
| 8. MASSETER | Raises lower jaw for chewing | Zygomatic arch | Mandible |
| 9. RISORIUS | Dimples corners of mouth | Hollow of cheek | Orbicularis oris at corner of mouth |
| 10. LEVATOR LABII SUPERIORIS | Elevates upper lip | Maxilla & zygomatic bone | Orbicularis oris & skin above lips |
| 11. BUCCINATOR | Compresses cheek | Maxilla | Orbicularis oris |
| 12. ORBICULARIS ORIS | Protrudes lips; closes and points mouth | Fascia around lips and muscles nearby | Lip-rim |
| 13. DEPRESSOR ANGULI ORIS | Brings corners of mouth downward | Mandible | Corner of mouth |
| 14. DEPRESSOR LABII INFERIORIS | Moves lower lip downward and outward | Mandible | Orbicularis oris |
| 15. MENTALIS | Elevates and protrudes lower lip | Mandible (chin) | Orbicularis oris |

All actions and descriptive terms are defined with reference to the ANATOMICAL POSITION, that is, a figure standing erect, feet parallel and flat on the floor, eyes directed forward, arms at the sides, and palms facing forward.

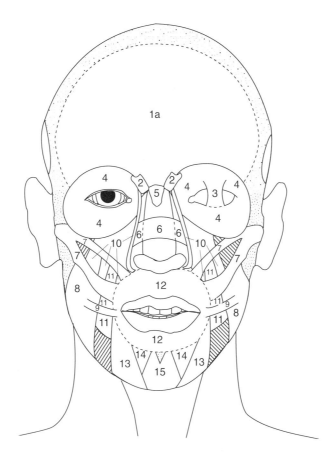

**Figure 21** *Stephen Rogers Peck. Coloring book head, anterior (front). Flayed head, anterior (front).* From An Atlas of Human Anatomy for the Artist *by Stephen Rogers Peck. Copyright © 1951 by Oxford University Press, Inc.; renewed 1979 by Stephen Rogers Peck. Reprinted by permission of Oxford University Press, Inc.*

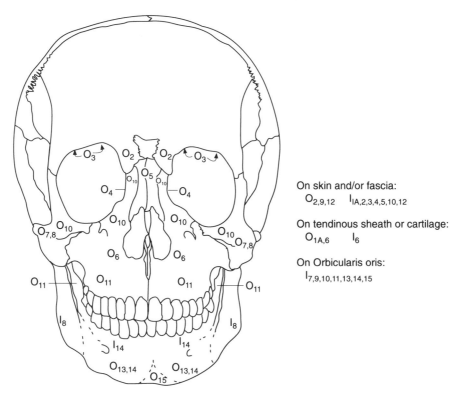

On skin and/or fascia:
$O_{2,9,12}$    $I_{IA,2,3,4,5,10,12}$

On tendinous sheath or cartilage:
$O_{1A,6}$    $I_6$

On Orbicularis oris:
$I_{7,9,10,11,13,14,15}$

**Figure 22** *Albinus (1697–1770). The skull, anterior (front).* [Albinus on Anatomy. *By Robert Beverly Hale and Terrance Coyle. Watson-Guptill Pub., NY. 1979. p. 54.]*

| Name | Action | Origin | Insertion |
|---|---|---|---|
| 1. EPICRANIUS:<br>(a) Frontal belly<br>(b) Occipital belly | (a) Moves scalp backwards<br>(b) Wrinkles forehead | (a) Sheath of occipital belly<br>(b) Occipital bone and mastoid process | (a) Skin of eyebrows<br>(b) Sheath of frontal belly |
| 2. CORRUGATOR | Draws eyebrows toward midline | Nasal part of frontal bone and tissue above eyebrow | Root of nose |
| 3. LEVATOR PALPEBRAE | Opens upper eyelid | Orbital roof | Skin of upper eyelid |
| 4. ORBICULARIS OCULI | Firmly closes eyelid | Rim of orbit | Edge of lidslit |
| 5. PROCERUS | Wrinkles skin horizontally at bridge of nose | Nasal bones | Skin in between eyebrows |
| 6. NASALIS | Compresses nostrils | Maxilla nasal cartilage | Sheath covering nostrils |
| 7. ZYGOMATICUS | Moves corners of mouth outward and upward | Zygomatic bone | Orbicularis oris and skin at corners of mouth |
| 8. MASSETER | Raises lower jaw for chewing | Zygomatic arch | Mandible |
| 9. RISORIUS | Dimples corners of mouth | Hollow of cheek | Orbicularis oris at corner of mouth |
| 10. LEVATOR LABII SUPERIORIS | Elevates upper lip | Maxilla & zygomatic bone | Orbicularis oris & skin above lips |
| 11. BUCCINATOR | Compresses cheek | Maxilla | Orbicularis oris |
| 12. ORBICULARIS ORIS | Protrudes lips; closes and points mouth | Fascia around lips and muscles nearby | Lip-rim |
| 13. DEPRESSOR ANGULI ORIS | Brings corners of mouth downward | Mandible | Corner of mouth |
| 14. DEPRESSOR LABII INFERIORIS | Moves lower lip downward and outward | Mandible | Orbicularis oris |
| 15. MENTALIS | Elevates and protrudes lower lip | Mandible (chin) | Orbicularis oris |

All actions and descriptive terms are defined with reference to the ANATOMICAL POSITION, that is, a figure standing erect, feet parallel and flat on the floor, eyes directed forward, arms at the sides, and palms facing forward.

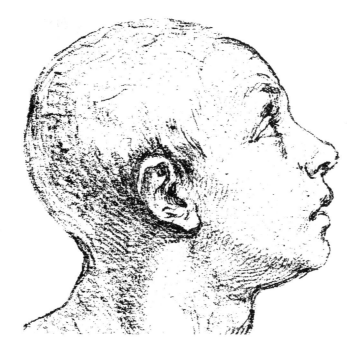

**Figure 23**  *Abraham Bloemart. Adapted from* Two Studies of the Head of a Young Man, and a Right Hand Holding an Open book. *(recto). Christ Church, Oxford England. [Seen in* Old Masters Drawings from The Governing Body, Christ Church, Oxford. *By James Byam Shaw. International Exhibitions Foundation, 1972–1973. Plate 91.]*

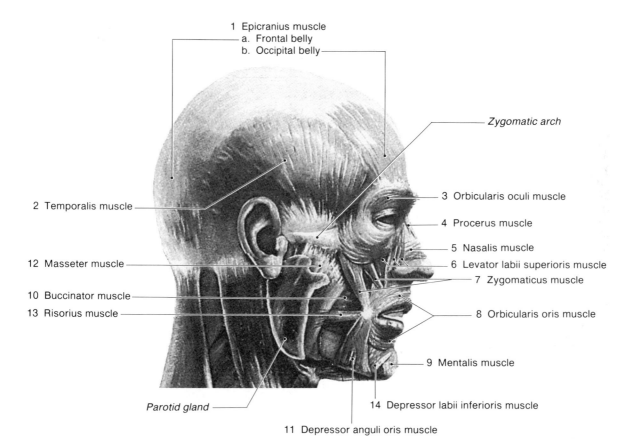

1 Epicranius muscle
  a. Frontal belly
  b. Occipital belly

*Zygomatic arch*

2 Temporalis muscle

3 Orbicularis oculi muscle

4 Procerus muscle

5 Nasalis muscle

12 Masseter muscle

6 Levator labii superioris muscle

7 Zygomaticus muscle

10 Buccinator muscle

13 Risorius muscle

8 Orbicularis oris muscle

9 Mentalis muscle

*Parotid gland*

14 Depressor labii inferioris muscle

11 Depressor anguli oris muscle

**Figure 24**  *Stephen Rogers Peck. Flayed head, lateral (side). From* An Atlas of Human Anatomy for the Artist *by Stephen Rogers Peck. Copyright © 1951 by Oxford University Press, Inc.; renewed 1979 by Stephen Rogers Peck. Reprinted by permission of Oxford University Press, Inc.*

| Name | Action | Origin | Insertion |
|---|---|---|---|
| 1. EPICRANIUS:<br>(a) Frontal belly<br>(b) Occipital belly | (a) Wrinkles forehead<br>(b) Moves scalp backwards | (a) Frontal belly:<br>Sheath of occipital belly<br>(b) Occipital belly:<br>Occipital bone and mastoid process | (a) Skin of eyebrows<br>(b) Sheath of frontal belly |
| 2. TEMPORALIS | Raises lower jaw for chewing and speaking | Temporal bone | Mandible |
| 3. ORBICULARIS OCULI | Firmly closes eyelid | Rim of orbit | Edge of lidslit |
| 4. PROCERUS | Wrinkles skin horizontally at bridge of nose | Nasal bones | Skin in between eyebrows |
| 5. NASALIS | Compresses nostrils | Maxilla & nasal cartilage | Sheath covering nostrils |
| 6. LEVATOR LABII SUPERIORIS | Elevates upper lip | Maxilla & zygomatic bone | Orbicularis oris & skin above lips |
| 7. ZYGOMATICUS | Moves corners of mouth outward and upward | Zygomatic bone | Orbicularis and skin at corners of mouth |
| 8. ORBICULARIS ORIS | Protrudes lips; closes and points mouth | Fascia around lips and muscles nearby | Lip-rim |
| 9. MENTALIS | Elevates and protrudes lower lip | Mandible (chin) | Orbicularis oris |
| 10. BUCCINATOR | Compresses cheek | Maxilla and mandible | Orbicularis oris |
| 11. DEPRESSOR ANGULI ORIS | Brings corners of mouth downward | Mandible | Corner of mouth |
| 12. MASSETER | Raises lower jaw for chewing | Zygomatic arch | Mandible |
| 13. RISORIUS | Dimples corners of mouth | Hollow of cheek | Orbicularis oris at corner of mouth |
| 14. DEPRESSOR LABII INFERIORIS | Moves lower lip downward and outward | Mandible | Orbicularis oris |

All actions and descriptive terms are defined with reference to the ANATOMICAL POSITION, that is, a figure standing erect, feet parallel and flat on the floor, eyes directed forward, arms at the sides, and palms facing forward.

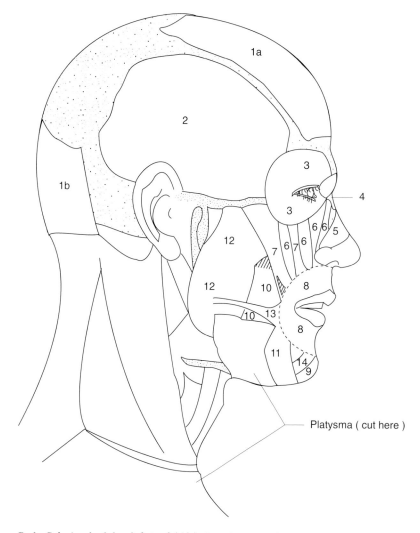

Platysma ( cut here )

**Figure 25** *Stephen Rogers Peck. Coloring book head, lateral (side). From* An Atlas of Human Anatomy for the Artist *by Stephen Rogers Peck. Copyright © 1951 by Oxford University Press, Inc.; renewed 1979 by Stephen Rogers Peck. Reprinted by permission of Oxford University Press, Inc.*

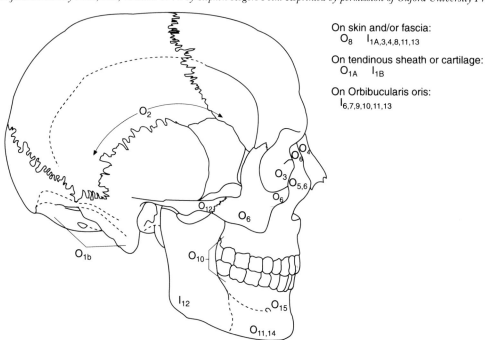

On skin and/or fascia:
$O_8$   $I_{1A,3,4,8,11,13}$

On tendinous sheath or cartilage:
$O_{1A}$   $I_{1B}$

On Orbibucularis oris:
$I_{6,7,9,10,11,13}$

**Figure 26** *Albinus (1697–1770). Skull, lateral (side). [Reversed tracing from Albinus. p. 58.]*

| Name | Action | Origin | Insertion |
|---|---|---|---|
| 1. EPICRANIUS:<br>(a) Frontal belly<br>(b) Occipital belly | (a) Wrinkles forehead<br>(b) Moves scalp backwards | (a) Frontal belly:<br>    Sheath of occipital belly<br>(b) Occipital belly:<br>    Occipital bone and mastoid<br>    process | (a) Skin of eyebrows<br>(b) Sheath of frontal belly |
| 2. TEMPORALIS | Raises lower jaw for chewing and speaking | Temporal bone | Mandible |
| 3. ORBICULARIS OCULI | Firmly closes eyelid | Rim of orbit | Edge of lidslit |
| 4. PROCERUS | Wrinkles skin horizontally at bridge of nose | Nasal bones | Skin in between eyebrows |
| 5. NASALIS | Compresses nostrils | Maxilla & nasal cartilage | Sheath covering nostrils |
| 6. LEVATOR LABII SUPERIORIS | Elevates upper lip | Maxilla & zygomatic bone | Orbicularis oris & skin above lips |
| 7. ZYGOMATICUS | Moves corners of mouth outward and upward | Zygomatic bone | Orbicularis and skin at corners of mouth |
| 8. ORBICULARIS ORIS | Protrudes lips; closes and points mouth | Fascia around lips and muscles nearby | Lip-rim |
| 9. MENTALIS | Elevates and protrudes lower lip | Mandible (chin) | Orbicularis oris |
| 10. BUCCINATOR | Compresses cheek | Maxilla and mandible | Orbicularis oris |
| 11. DEPRESSOR ANGULI ORIS | Brings corners of mouth downward | Mandible | Corner of mouth |
| 12. MASSETER | Raises lower jaw for chewing | Zygomatic arch | Mandible |
| 13. RISORIUS | Dimples corners of mouth | Hollow of cheek | Orbicularis oris at corner of mouth |
| 14. DEPRESSOR LABII INFERIORIS | Moves lower lip downward and outward | Mandible | Orbicularis oris |

All actions and descriptive terms are defined with reference to the
ANATOMICAL POSITION, that is, a figure standing erect, feet
parallel and flat on the floor, eyes directed forward, arms at the sides,
and palms facing forward.

**Figure 27** *Student drawing after Leonardo da Vinci (1452–1519).* Nude Man Standing, Back to Spectator, *red chalk, 10 5/8" ×
6 5/16" (270 × 160 mm). Windsor Castle, Royal Library. © 1991 Her Majesty. Queen Elizabeth II.*

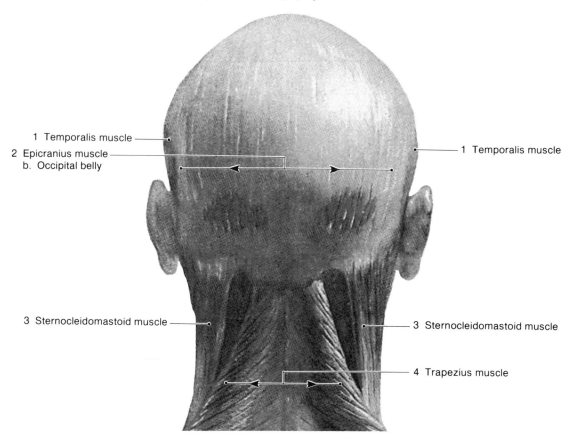

1 Temporalis muscle
2 Epicranius muscle
  b. Occipital belly

1 Temporalis muscle

3 Sternocleidomastoid muscle

3 Sternocleidomastoid muscle

4 Trapezius muscle

**Figure 28** *Stephen Rogers Peck. Flayed head, posterior (back). From* An Atlas of Human Anatomy for the Artist *by Stephen Rogers Peck. Copyright
© 1951 by Oxford University Press, Inc.; renewed 1979 by Stephen Rogers Peck. Reprinted by permission of Oxford University Press, Inc.*

| Name | Action | Origin | Insertion |
|------|--------|--------|-----------|
| 1. TEMPORALIS | Raises lower jaw for chewing and speaking | Temporal bone | Mandible |
| 2. EPICRANIUS:<br>(a) Frontal belly<br>(b) Occipital belly | (a) Wrinkles forehead<br>(b) Moves scalp backwards | (a) Sheath of occipital belly<br>(b) Occipital belly: Occipital bone and mastoid process | (a) Skin of eyebrows<br>(b) Sheath of frontal belly |
| 3. STERNOCLEIDOMASTOID | Lifts head; tips head backward; turns head to side | Sternum & clavicle | Mastoid process of the temporal bone of the skull |
| 4. TRAPEZIUS | Brings scapulae together; moves them up and down; draws head backwards | Occipital bone (on skull) down to thoracic vertebra #12 | Clavicle, acromion, & spine of the scapula |

All actions and descriptive terms are defined with reference to the ANATOMICAL POSITION, that is, a figure standing erect, feet parallel and flat on the floor, eyes directed forward, arms at the sides, and palms facing forward.

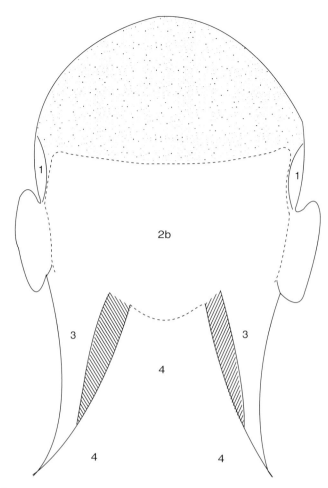

**Figure 29** *Stephen Rogers Peck. Coloring book head, posterior (back).* From An Atlas of Human Anatomy for the Artist *by Stephen Rogers Peck. Copyright © 1951 by Oxford University Press, Inc.; renewed 1979 by Stephen Rogers Peck. Reprinted by permission of Oxford University Press, Inc.*

On fascia or tendinous sheath:
$I_2$

On mandible:
$I_1$

On sternum or clavicle:
$O_3$

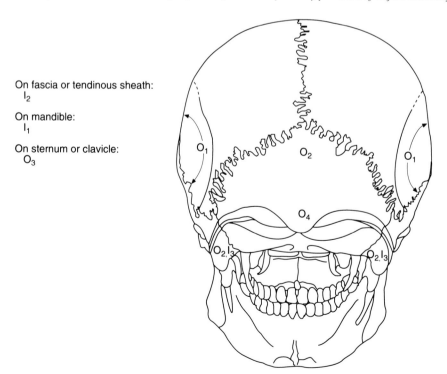

**Figure 30** *Albinus (1697–1770). Skull, posterior (back).* [Albinus on Anatomy. By Robert Beverly Hale and Terrance Coyle. Watson-Guptill Pub., NY. 1979. p. 56.]

| Name | Action | Origin | Insertion |
|------|--------|--------|-----------|
| 1. TEMPORALIS | Raises lower jaw for chewing and speaking | Temporal bone | Mandible |
| 2. EPICRANIUS:<br>(a) Frontal belly<br>(b) Occipital belly | (a) Wrinkles forehead<br>(b) Moves scalp backwards | (a) Sheath of occipital belly<br>(b) Occipital belly: Occipital bone and mastoid process | (a) Skin of eyebrows<br>(b) Sheath of frontal belly |
| 3. STERNOCLEIDOMASTOID | Lifts head; tips head backward; turns head to side | Sternum & clavicle | Mastoid process of the temporal bone of the skull |
| 4. TRAPEZIUS | Brings scapulae together; moves them up and down; draws head backwards | Occipital bone (on skull) down to thoracic vertebra #12 | Clavicle, acromion, & spine of the scapula |

All actions and descriptive terms are defined with reference to the ANATOMICAL POSITION, that is, a figure standing erect, feet parallel and flat on the floor, eyes directed forward, arms at the sides, and palms facing forward.

## Skull

Once again we move to *Sensitizing* with blind contour, *Structuring* with sighting and volume, and *Superimposing*, bones into flesh, this time using the skull.

After placing yourself somewhere close to the skull, begin with a clean sheet of drawing paper. Remember to "draw under cover." Look at the whole shape of the skull and try to see the subtle undulations of the line, the "edges" of the form of that head, now bone. Progress slowly, inching your pencil along, never looking at what you draw. Do not move your eye ahead of your pencil. Slowly and easily let your pencil move, pressing down for portions of the skull that appear to move away from you, less on the lines when portions move toward you. Move into sections of the face, the sockets, the teeth, the mandibles and such, all the while continuing your work with the contour line. Do not look at your paper. Reposition your pencil when you need to. Accuracy in this exercise is not important. Sensitivity is. Let your surprise and satisfaction be in the character of the line. You may find a humorous end result for a skeleton image.

The second exercise, *Structure*, uses gesture and sighting and volumes. Take another sheet of drawing paper and a sharpened pencil and begin sighting, referring now to your paper. The skull is a somewhat big and bulky unit. Most students like to linger on the contour line, the edge. This exercise, structure, helps ease you off that comforting shape.

Let's begin with lines that suggest which way the skull is positioned on the neck. Does the skull tilt left or right? The skull probably tilts one way or another. One line, slightly diagonal will do it. What kind of single directional line is indicated by the tops of the pair of eye sockets? If the head is at angle to the neck, so the sockets will be. That means another line. Look at the nasal bones. What angle is that opening? What single line would you use to indicate the nasal negative space to the neck? Another line, probably diagonal, too. What kind of line informs the mass of the teeth? A curvilinear line? Try to think line, not object.

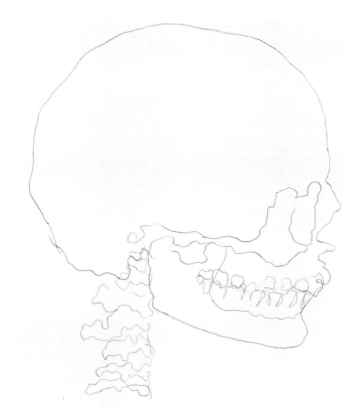

**Figure 31** *Blind contour of a skull.*

Having established lines suggesting teeth, now draw the rest of the skull in directional lines. You may be referring to and drawing *some* edges. But SEE the edges as ANGLES. Try NOT to draw the object.

Just a cautionary note: if you outline the head again while looking at it, and have simply redone the contour, you will not learn much, you will just rework old habits. Gesture is sighting one thing to another. Use lines that indicate those reference points previously mentioned—such as brow to other brow, but do not draw the item in contour.

**Figure 32** *Sighting, gesture, volumetric drawing of a skull.*

Now look for volumes in the skull and decide what the most appropriate geometric shape for that volume would be. Is it an oval, an egg, a sphere, a cube? What does the large skull shape appear to be?

Pencil to paper and begin. If you are worried about accuracy here, ease your mind a little and just look at the skull, letting your pencil carry itself all the way around the head, vertically and horizontally, making something that will probably resemble a large egg. That image should help get you going. Drawing a gesture of one large object is confusing because so much of the information is packed more obviously in the edges creating boundaries. Rethinking your drawing methods helps you see one large shape.

When the large geometric shape is finished, move into the smaller volumes such as the zygomatic arch, the mandibles, and such.

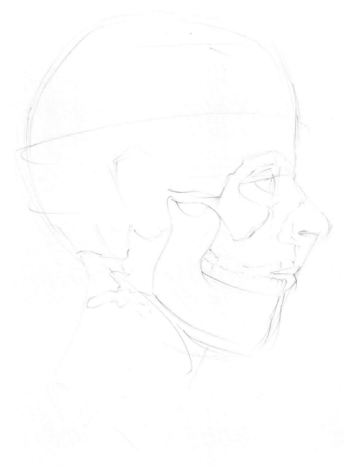

**Figure 33** *Bones superimposed into flesh.*

And now the last of the three exercises, the *Superimposition*. Begin with a gesture of your model's head, some careful but light, lined work with volumes throughout the face. Then draw the external contour line of the head. When that preliminary drawing is done, refer to the skull and begin placing bones into the flesh drawing of the head you just completed of the model. If you become confused about the placement of bone to tissue, press your fingers to your own face to feel bone positions. Refer to illustrations in the book as well. ▲

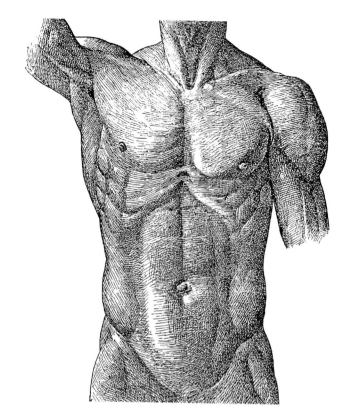

**Figure 34**  *Surface torso, anterior (front). As appeared in* An Atlas of Anatomy for Artists. *By Fritz Schider. Dover Publications, Inc. NY.*

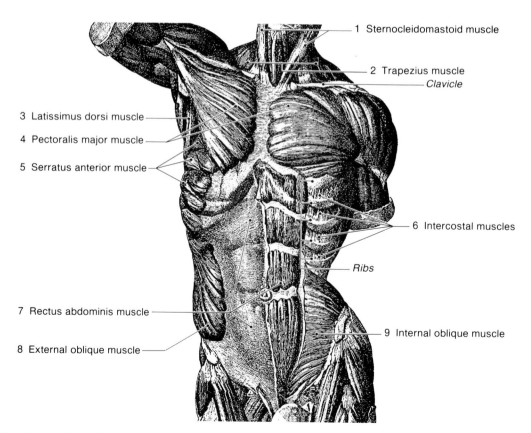

1 Sternocleidomastoid muscle

2 Trapezius muscle
— *Clavicle*

3 Latissimus dorsi muscle

4 Pectoralis major muscle

5 Serratus anterior muscle

6 Intercostal muscles

*Ribs*

7 Rectus abdominis muscle

8 External oblique muscle

9 Internal oblique muscle

**Figure 35**  *Flayed torso, anterior (front). As appeared in* An Atlas of Anatomy for Artists. *By Fritz Schider. Dover Publications, Inc. NY.*

36

| Name | Action | Origin | Insertion |
|------|--------|--------|-----------|
| 1. STERNOCLEIDOMASTOID | Lifts head; tips head backward; turns head to side | Sternum & clavicle | Mastoid process of the temporal bone of skull |
| 2. TRAPEZIUS | Brings scapulae together; moves them up and down; draws head backwards | Occipital bone (on skull) down to thoracic vertebrae #12 | Clavicle, acromion, & spine of scapula |
| 3. LATISSIMUS DORSI | Moves arm backward and toward spine; rotates arm inward; lowers arm from vertical position | Spines of thoracic, lumbar and sacral vertebrae, lower ribs, and iliac crest | Scapula |
| 4. PECTORALIS MAJOR | Moves arm forward and across body; rotates arm inward; lowers arm from vertical position | Clavicle, sternum, & cartilage between ribs #2–6 | Humerus |
| 5. SERRATUS ANTERIOR | Pulls scapulae downward and forward | The 8 to 9 uppermost ribs | Scapula |
| 6. INTERCOSTALS | Moves ribs during breathing | Bottom border of rib | Top border of the next rib down |
| 7. RECTUS ABDOMINIS | Flexes spine | Pubic crest and symphysis pubis | Ribs #5–7 & xiphoid process on sternum |
| 8. EXTERNAL OBLIQUE | Works with rectus abdominus to flex and twist spine | Lower eight ribs | Iliac crest and sheath covering RECTUS ABDOMINUS |
| 9. INTERNAL OBLIQUE | Same as external oblique | Iliac crest and groin area | Lower three ribs |

All actions and descriptive terms are defined with reference to the ANATOMICAL POSITION, that is, a figure standing erect, feet parallel and flat on the floor, eyes directed forward, arms at the sides, and palms facing forward.

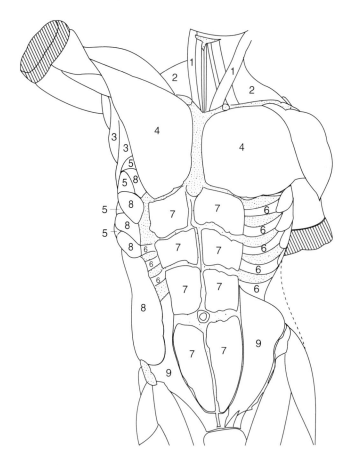

**Figure 36** *Coloring book of torso, anterior (front). As appeared in* An Atlas of Anatomy for Artists. *By Fritz Schider, Dover Publications, Inc. N.Y.*

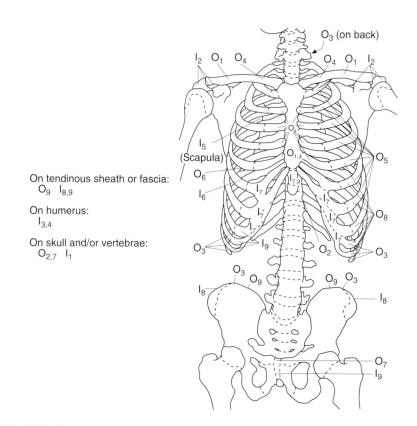

$O_3$ (on back)

$I_2$ $O_1$ $O_4$     $O_4$ $O_1$ $I_2$

$I_5$
(Scapula)

$O_{1,4}$

$O_5$

On tendinous sheath or fascia:
  $O_9$  $I_{8,9}$

$O_6$

$I_6$

$I_{7,9}$

On humerus:
  $I_{3,4}$

$I_7$    $I_7$

$O_8$

On skull and/or vertebrae:
  $O_{2,7}$  $I_1$

$O_3$     $I_9$     $O_2$    $O_3$

$O_3$ $O_9$    $O_9$ $O_3$

$I_8$            $I_8$

$O_7$

$I_9$

**Figure 37** *Albinus (1697–1770). Torso bones, anterior (front). [*Albinus on Anatomy. *By Robert Beverly Hale and Terrance Coyle. Watson-Guptill Pub., NY. 1979. p. 28.]*

| Name | Action | Origin | Insertion |
|---|---|---|---|
| 1. STERNOCLEIDOMASTOID | Lifts head; tips head backward; turns head to side | Sternum & clavicle | Mastoid process of the temporal bone of skull |
| 2. TRAPEZIUS | Brings scapulae together; moves them up and down; draws head backwards | Occipital bone (on skull) down to thoracic vertebrae #12 | Clavicle, acromion, & spine of scapula |
| 3. LATISSIMUS DORSI | Moves arm backward and toward spine; rotates arm inward; lowers arm from vertical position | Spines of thoracic, lumbar and sacral vertebrae, lower ribs, and iliac crest | Scapula |
| 4. PECTORALIS MAJOR | Moves arm forward and across body; rotates arm inward; lowers arm from vertical position | Clavicle, sternum, & cartilage between ribs #2–6 | Humerus |
| 5. SERRATUS ANTERIOR | Pulls scapulae downward and forward | The 8 to 9 uppermost ribs | Scapula |
| 6. INTERCOSTALS | Moves ribs during breathing | Bottom border of rib | Top border of the next rib down |
| 7. RECTUS ABDOMINIS | Flexes spine | Pubic crest and symphysis pubis | Ribs #5–7 & xiphoid process on sternum |
| 8. EXTERNAL OBLIQUE | Works with rectus abdominus to flex and twist spine | Lower eight ribs | Iliac crest and sheath covering RECTUS ABDOMINUS |
| 9. INTERNAL OBLIQUE | Same as external oblique | Iliac crest and groin area | Lower three ribs |

All actions and descriptive terms are defined with reference to the ANATOMICAL POSITION, that is, a figure standing erect, feet parallel and flat on the floor, eyes directed forward, arms at the sides, and palms facing forward.

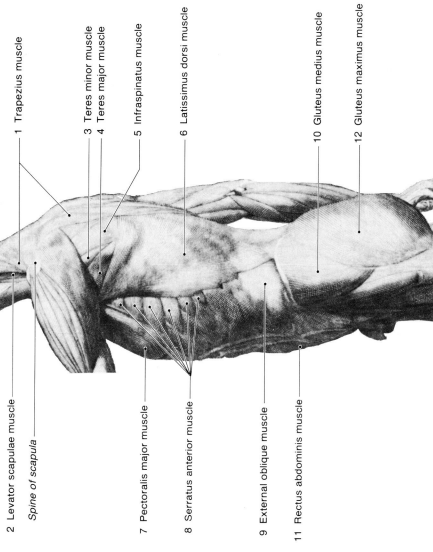

1  Trapezius muscle
3  Teres minor muscle
4  Teres major muscle
5  Infraspinatus muscle
6  Latissimus dorsi muscle
10  Gluteus medius muscle
12  Gluteus maximus muscle

2  Levator scapulae muscle
Spine of scapula

7  Pectoralis major muscle
8  Serratus anterior muscle
9  External oblique muscle
11  Rectus abdominis muscle

**Figure 39**  *Albinus (1697–1770). Flayed torso, lateral (side). [Albinus on Anatomy. By Robert Beverly Hale and Terrance Coyle. Watson-Guptill Pub., NY. 1979. p. 39.]*

**Figure 38**  *Surface torso, lateral (side) (lithograph). (Courtesy of the Francis A. Countway Library of Medicine.)*

| Name | Action | Origin | Insertion |
|------|--------|--------|-----------|
| 1. TRAPEZIUS | Brings scapulae together; moves them up and down; draws head backward | Occipital bone (on skull) down to thoracic vertebra #12 | Clavicle, acromion, & spine of scapula |
| 2. LEVATOR SCAPULAE | Moves scapulae together and upward | Fourth and fifth cervical vertebrae | Scapula |
| 3. TERES MINOR | Brings outstretched arm toward body; rotates arm outward | Scapula, near underarm | Humerus |
| 4. TERES MAJOR | Brings outstretched arm toward body; rotates arm inward; lowers arm from vertical position | Lower part of scapula | Humerus |
| 5. INFRASPINATUS | Fully rotates arm outward; helps to raise arm above head | Scapula | Humerus |
| 6. LATISSIMUS DORSI | Moves arm backward and toward spine; rotates arm inward; lowers arm from vertical position | Spines of thoracic, lumbar, and sacral vertebrae, lower ribs, and iliac crest | Humerus |
| 7. PECTORALIS MAJOR | Moves arm forward and across body; rotates arm inward; lowers arm from vertical position | Clavicle, sternum, & cartilage between ribs #2–6 | Humerus |
| 8. SERRATUS ANTERIOR | Pulls scapulae downward and forward | The 8 to 9 uppermost ribs | Scapula |
| 9. EXTERNAL OBLIQUE | Works with rectus abdominus to flex and twist spine | Lower eight ribs | Iliac crest and sheath covering RECTUS ABDOMINUS |
| 10. GLUTEUS MEDIUS | Moves leg from standing position to outstretched side position | Ilium | Femur |
| 11. RECTUS ABDOMINIS | Flexes spine | Pubic crest and symphysis pubis | Ribs #5–7 & xiphoid process on sternum |
| 12. GLUTEUS MAXIMUS | Moves thigh backward; returns leg from side to standing position; rotates leg outward; presses buttocks together | Ilium crest, sacrum & coccyx | Shaft of femur |

All actions and descriptive terms are defined with reference to the
ANATOMICAL POSITION, that is, a figure standing erect, feet
parallel and flat on the floor, eyes directed forward, arms at the sides,
and palms facing forward.

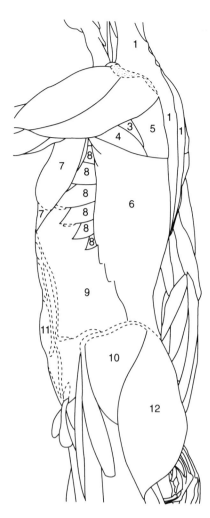

**Figure 40** Albinus (1697–1770). Coloring book of torso, lateral (side). [Albinus on Anatomy. *By Robert Beverly Hale and Terrance Coyle. Watson-Guptill Pub., NY. 1979. p. 38.]*

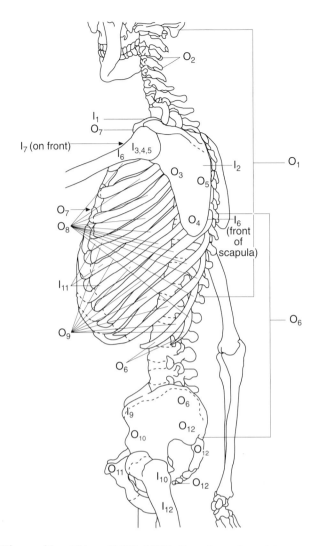

**Figure 41** Albinus (1697–1770). Torso bones, lateral (side). [Albinus on Anatomy. *By Robert Beverly Hale and Terrance Coyle. Watson-Guptill Pub., NY. 1979. p. 32.]*

| Name | Action | Origin | Insertion |
|------|--------|--------|-----------|
| 1. TRAPEZIUS | Brings scapulae together; moves them up and down; draws head backward | Occipital bone (on skull) down to thoracic vertebra #12 | Clavicle, acromion, & spine of scapula |
| 2. LEVATOR SCAPULAE | Moves scapulae together and upward | Fourth and fifth cervical vertebrae | Scapula |
| 3. TERES MINOR | Brings outstretched arm toward body; rotates arm outward | Scapula, near underarm | Humerus |
| 4. TERES MAJOR | Brings outstretched arm toward body; rotates arm inward; lowers arm from vertical position | Lower part of scapula | Humerus |
| 5. INFRASPINATUS | Fully rotates arm outward; helps to raise arm above head | Scapula | Humerus |
| 6. LATISSIMUS DORSI | Moves arm backward and toward spine; rotates arm inward; lowers arm from vertical position | Spines of thoracic, lumbar, and sacral vertebrae, lower ribs, and iliac crest | Humerus |
| 7. PECTORALIS MAJOR | Moves arm forward and across body; rotates arm inward; lowers arm from vertical position | Clavicle, sternum, & cartilage between ribs #2–6 | Humerus |
| 8. SERRATUS ANTERIOR | Pulls scapulae downward and forward | The 8 to 9 uppermost ribs | Scapula |
| 9. EXTERNAL OBLIQUE | Works with rectus abdominus to flex and twist spine | Lower eight ribs | Iliac crest and sheath covering RECTUS ABDOMINUS |
| 10. GLUTEUS MEDIUS | Moves leg from standing position to outstretched side position | Ilium | Femur |
| 11. RECTUS ABDOMINIS | Flexes spine | Pubic crest and symphysis pubis | Ribs #5–7 & xiphoid process on sternum |
| 12. GLUTEUS MAXIMUS | Moves thigh backward; returns leg from side to standing position; rotates leg outward; presses buttocks together | Ilium crest, sacrum & coccyx | Shaft of femur |

All actions and descriptive terms are defined with reference to the ANATOMICAL POSITION, that is, a figure standing erect, feet parallel and flat on the floor, eyes directed forward, arms at the sides, and palms facing forward.

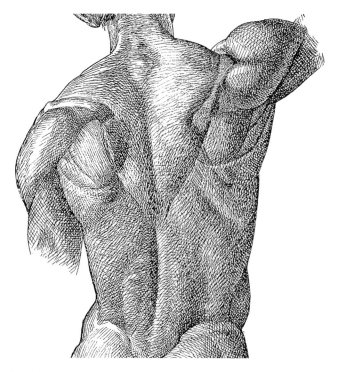

*Figure 42*    *Surface torso, posterior (back). As appeared in* An Atlas of Anatomy for Artists. *By Fritz Schider. Dover Publications, Inc. NY.*

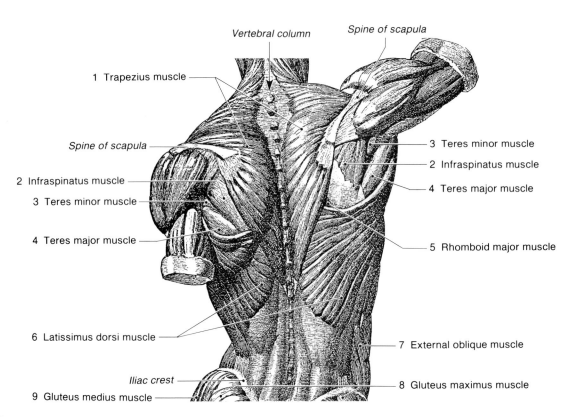

Vertebral column

Spine of scapula

1 Trapezius muscle

Spine of scapula

2 Infraspinatus muscle

3 Teres minor muscle

4 Teres major muscle

3 Teres minor muscle

2 Infraspinatus muscle

4 Teres major muscle

5 Rhomboid major muscle

6 Latissimus dorsi muscle

7 External oblique muscle

Iliac crest

9 Gluteus medius muscle

8 Gluteus maximus muscle

*Figure 43*    *Flayed torso, posterior (back). As appeared in* An Atlas of Anatomy for Artists. *By Fritz Schider. Dover Publications, Inc. NY.*

| Name | Action | Origin | Insertion |
|---|---|---|---|
| 1. TRAPEZIUS | Brings scapulae together; moves them up and down; draws head backward | Occipital bone (on skull) down to thoracic vertebrae #12 | Clavicle, acromion, & spine of scapula |
| 2. INFRASPINATUS | Fully rotates arm outward; helps to raise arm above head | Scapula | Humerus |
| 3. TERES MINOR | Brings outstretched arm toward body; rotates arm outward | Scapula, near underarm | Humerus |
| 4. TERES MAJOR | Brings outstretched arm toward body; rotates arm inward; lowers arm from vertical position | Lower part of scapula | Humerus |
| 5. RHOMBOID MAJOR | Moves scapulae toward each other and upward | Thoracic vertebrae #2–5 | Scapula |
| 6. LATISSIMUS DORSI | Moves arm backward and toward spine; rotates arm inward; lowers arm from vertical position | Spines of thoracic, lumbar, and sacral vertebrae, lower ribs, and iliac crest | Humerus |
| 7. EXTERNAL OBLIQUE | Works with rectus abdominus to flex and twist spine | Lower eight ribs | Iliac crest and sheath covering RECTUS ABDOMINUS |
| 8. GLUTEUS MAXIMUS | Moves thigh backward; returns leg from side to standing position; rotates leg outward; presses buttocks together | Iliac crest, sacrum, & coccyx | Shaft of femur |
| 9. GLUTEUS MEDIUS | Moves leg from standing position to outstretched side position | Ilium | Femur |

All actions and descriptive terms are defined with reference to the
ANATOMICAL POSITION, that is, a figure standing erect, feet
parallel and flat on the floor, eyes directed forward, arms at the sides,
and palms facing forward.

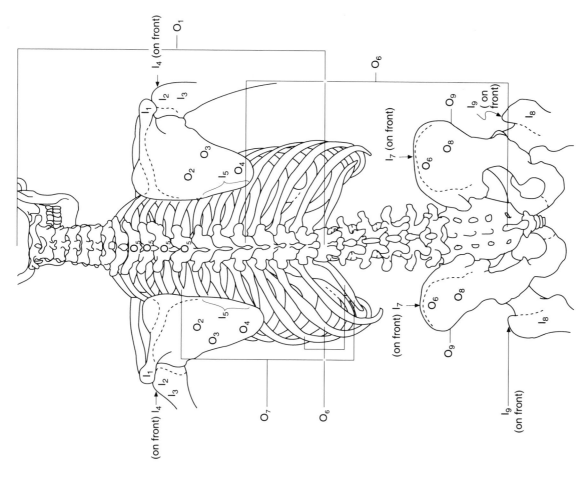

**Figure 45** *Albinus (1697–1770). Torso bones, posterior (back). [Albinus on Anatomy. By Robert Beverly Hale and Terrance Coyle. Watson-Guptill Pub., NY. 1979. p. 30.]*

**Figure 44** *Coloring book torso, posterior (back). As appeared in An Atlas of Anatomy for Artists. By Fritz Schider. Dover Publications, Inc. NY.*

| Name | Action | Origin | Insertion |
|------|--------|--------|-----------|
| 1. TRAPEZIUS | Brings scapulae together; moves them up and down; draws head backward | Occipital bone (on skull) down to thoracic vertebrae #12 | Clavicle, acromion, & spine of scapula |
| 2. INFRASPINATUS | Fully rotates arm outward; helps to raise arm above head | Scapula | Humerus |
| 3. TERES MINOR | Brings outstretched arm toward body; rotates arm outward | Scapula, near underarm | Humerus |
| 4. TERES MAJOR | Brings outstretched arm toward body; rotates arm inward; lowers arm from vertical position | Lower part of scapula | Humerus |
| 5. RHOMBOID MAJOR | Moves scapulae toward each other and upward | Thoracic vertebrae #2–5 | Scapula |
| 6. LATISSIMUS DORSI | Moves arm backward and toward spine; rotates arm inward; lowers arm from vertical position | Spines of thoracic, lumbar, and sacral vertebrae, lower ribs, and iliac crest | Humerus |
| 7. EXTERNAL OBLIQUE | Works with rectus abdominus to flex and twist spine | Lower eight ribs | Iliac crest and sheath covering RECTUS ABDOMINUS |
| 8. GLUTEUS MAXIMUS | Moves thigh backward; returns leg from side to standing position; rotates leg outward; presses buttocks together | Iliac crest, sacrum, & coccyx | Shaft of femur |
| 9. GLUTEUS MEDIUS | Moves leg from standing position to outstretched side position | Ilium | Femur |

All actions and descriptive terms are defined with reference to the ANATOMICAL POSITION, that is, a figure standing erect, feet parallel and flat on the floor, eyes directed forward, arms at the sides, and palms facing forward.

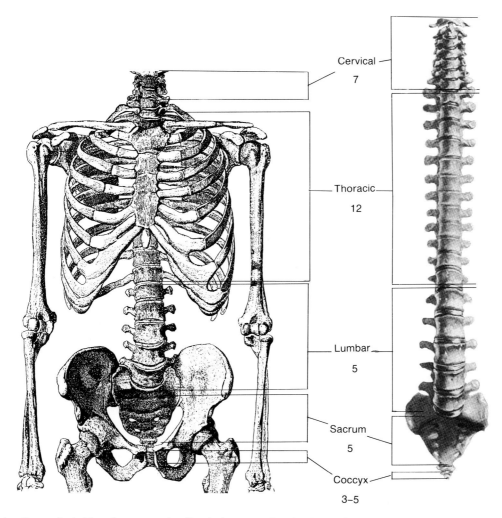

Cervical
7

Thoracic
12

Lumbar
5

Sacrum
5

Coccyx
3–5

**Figure 46**  *Stephen Rogers Peck. Torso bones, anterior (front). As appeared in* An Atlas of Anatomy for Artists. *By Fritz Schider. Dover Publications, Inc. NY. Spinal cord, anterior (front). 46B. From* An Atlas of Human Anatomy for the Artist *by Stephen Rogers Peck. Copyright © 1951 by Oxford University Press, Inc.; renewed 1979 by Stephen Rogers Peck. Reprinted by permission of Oxford University Press, Inc.*

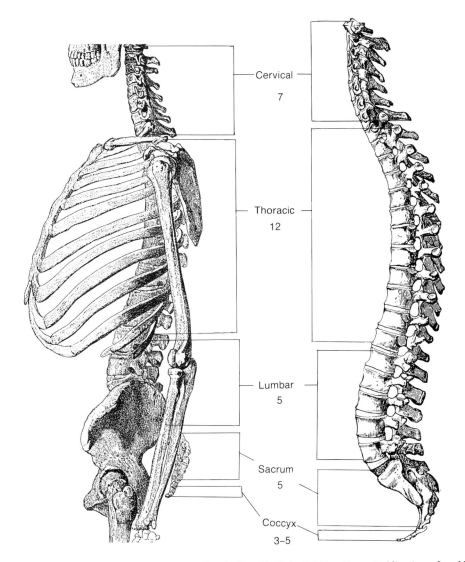

Cervical
7

Thoracic
12

Lumbar
5

Sacrum
5

Coccyx
3–5

**Figure 47** Torso bones, lateral (side). [An Atlas of Anatomy for Artists. *By Fritz Schider. Dover Publications, Inc. NY. 1957. pl. 10.*]
Spinal cord, lateral (side). [Gray's Anatomy. Bounty Books, NY, Fifteenth English Edition, p. 51, figure #22.]

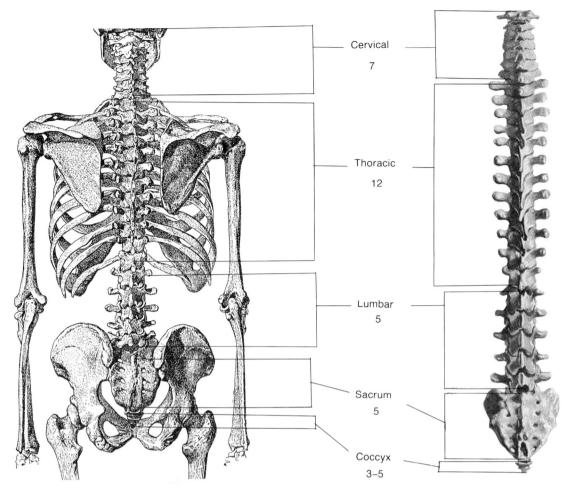

Cervical
7

Thoracic
12

Lumbar
5

Sacrum
5

Coccyx
3–5

**Figure 48** *Torso bones, posterior (back). As appeared in* An Atlas of Anatomy for Artists. *By Fritz Schider. Dover Publications, Inc. NY. Spinal cord, posterior (back). 48B. From* An Atlas of Human Anatomy for the Artist *by Stephen Rogers Peck. Copyright © 1951 by Oxford University Press, Inc.; renewed 1979 by Stephen Rogers Peck. Reprinted by permission of Oxford University Press, Inc.*

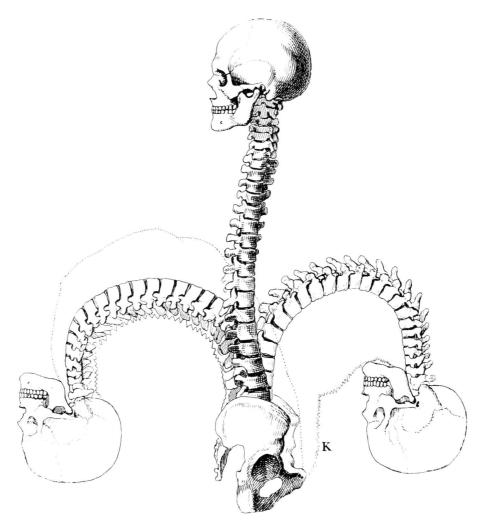

**Figure 49** *Jean Galbert Savage. Movements of the vertebral column. Left: Extension Right: Flexion. Courtesy of the Francis A. Countway Library of Medicine.*

## Trunk

The threefold sequence of exercises we have spoken of in this section of the book can be produced either from the front, side, back or three-quarter view. You choose. Whenever you begin with the blind contour, try as you draw to name the bones. If naming bones as you work is too distracting, then name each bone first before you begin to draw. Once seated or standing at your table or easel, cover your drawing hand with another sheet of drawing paper and begin the blind contour of the torso. Again, your attention should be exclusively on the apparent edge of the bone form. Look for distinctive differences in the undulations of the bones. Some of the bones indent, some do not. Keep your pencil moving carefully with the movement of your eye. Doing this exercise half-heartedly will weaken your line, yielding less information in the drawing. Information, of course, here refers to the *sensitivity of the line* rather than to a realistic drawing. There is a big difference.

A fresh sheet of paper is needed for the next problem, the torso *structure*. Without the head for a standard, or scale, you will need to find another module for consistent length; the sternum, for instance. Draw the large and generalized lines of direction, that is, the posture line of the rib cage/hip, the sternum, the clavicles, the angle of the hips. Draw the ribs carefully. A tendency with many students is to overlook the vertebral origin of each rib. Two extra ribs "float," attaching only to the back bone. The norm is 12 pairs of ribs; however not all skeletons will have this number and sometimes ribs are fused together. At any rate, look to your specific skeleton and take care that you concentrate on angles, one to another, heights and widths AND the spaces between the ribs. Generally you will see not only the negative space, but the ribs on the farthest side between the space, which need to be drawn. Drawing ribs is much easier if you keep in mind the volume of the cage. Think of the bones going "around" when you draw the torso in this structure exercise. Otherwise you may get lost, disoriented, or very discouraged.

Once the sighting with lines is drawn, go back carefully to look at the volumes of the bones. Using cross contour and geometric shapes, fill the lines into volumes. Geometric volumes should be used again—cones, cylinders,

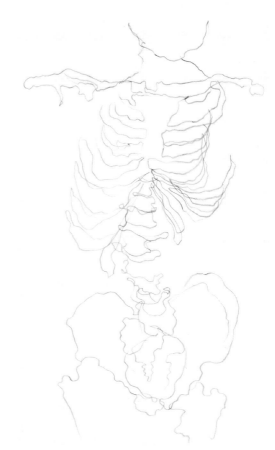

***Figure 50*** *Blind contour of the torso.*

cubes, whatever. The actual bones will be drawn as bones in the next exercise. Here we just look at their volume.

For the final exercise, the *superimposition*, draw the model's torso first, sighting, gestures, and volumes. Again, you now may have to adjust length and width of bone to superimpose them into the trunk from your point of view. And, as you think/draw volumes, insinuate the shape of each bone into the flesh. Your drawings will be much richer in information if you think sculpturally, meaning if you think of the form as though you could see it all the way around.  ▲

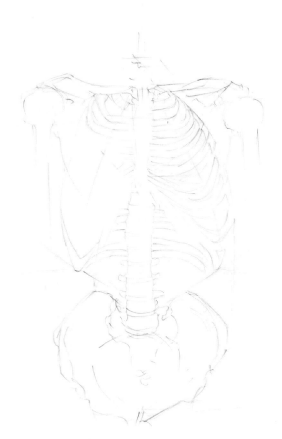

**Figure 51** *A structural drawing of the torso.*

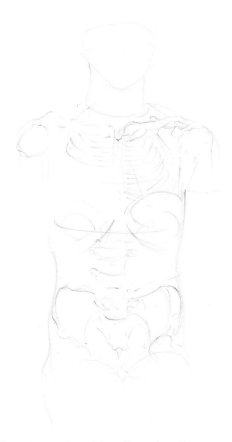

**Figure 52** *The superimposition of bones into the torso.*

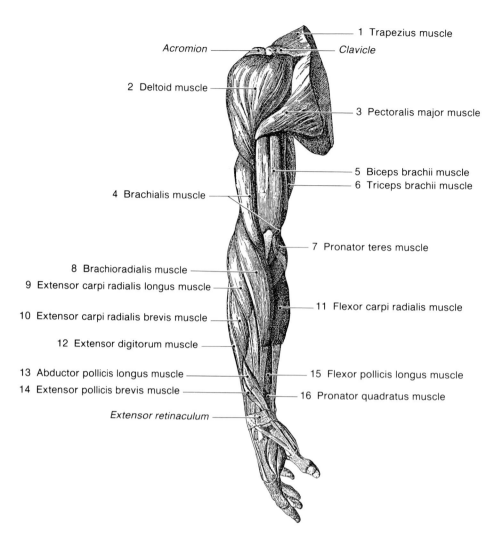

1 Trapezius muscle

Acromion ——— ——— Clavicle

2 Deltoid muscle

3 Pectoralis major muscle

5 Biceps brachii muscle
6 Triceps brachii muscle

4 Brachialis muscle

7 Pronator teres muscle

8 Brachioradialis muscle
9 Extensor carpi radialis longus muscle

10 Extensor carpi radialis brevis muscle

11 Flexor carpi radialis muscle

12 Extensor digitorum muscle

13 Abductor pollicis longus muscle
14 Extensor pollicis brevis muscle

15 Flexor pollicis longus muscle
16 Pronator quadratus muscle

Extensor retinaculum

**Figure 53** *Surface arm, anterior (front). As appeared in* An Atlas of Anatomy for Artists. *By Fritz Schider. Dover Publications, Inc. NY.*

**Figure 54** *Flayed arm, anterior (front). As appeared in* An Atlas of Anatomy for Artists. *By Fritz Schider. Dover Publications, Inc. NY.*

| Name | Action | Origin | Insertion |
|------|--------|--------|-----------|
| 1. TRAPEZIUS | Brings scapulae together; moves them up and down; draws head backward | Occipital bone (on skull) down to thoracic vertebra #12 | Clavicle, acromion, & spine of scapula |
| 2. DELTOID | Moves arm forward and backward; moves arm to side away from body | Clavicle, acromion, & spine of scapula | Humerus |
| 3. PECTORALIS MAJOR | Moves arm forward and across body; rotates arm inward; lowers arm from vertical position | Clavicle, sternum, & spine of scapula | Humerus |
| 4. BRACHIALIS | Bends arm at elbow | Humerus | Ulna (inside elbow joint) |
| 5. BICEPS BRACHII | Bends arm at elbow | Coracoid process of scapula | Radius |
| 6. TRICEPS BRACHII | Straightens elbow | Humerus & scapula | Ulna (on olecranon) |
| 7. PRONATOR TERES | Pronates forearm | Humerus | Radius |
| 8. BRACHIORADIALIS | Bends arm at elbow; while arm is bent, pronates forearm; when arm is straight, supinates forearm | Humerus | Radius |
| 9. EXTENSOR CARPI RADIALIS LONGUS | Same as brachioradialis; plus straightens wrist and moves hand outward from side of body at wrist joint | Humerus | Metacarpal #2 |
| 10. EXTENSOR CARPI RADIALIS BREVIS | Straightens wrist; moves hand outward from side of body at wrist joint | Humerus | Metacarpal #3 |
| 11. FLEXOR CARPI RADIALIS | Bends wrist; moves hand toward side of body at wrist joint | Humerus | Metacarpals #2 & #3 |
| 12. EXTENSOR DIGITORUM | Straightens fingers & hand; spreads apart fingers | Humerus | By tendons to phalanges #2–5 |
| 13. ABDUCTOR POLLICIS LONGUS | Moves thumb out from side of hand; moves thumb forward | Radius & ulna | Metacarpal #1 |
| 14. EXTENSOR POLLICIS BREVIS | Straightens final two thumb joints | Radius | First phalanx of thumb |
| 15. FLEXOR POLLICIS LONGUS | Flexes thumb | Radius & ulna | Second phalanx of thumb |
| 16. PRONATOR QUADRATUS | Pronates hands | Lower ulna | Lower radius |

All actions and descriptive terms are defined with reference to the ANATOMICAL POSITION, that is, a figure standing erect, feet parallel and flat on the floor, eyes directed forward, arms at the sides, and palms facing forward.

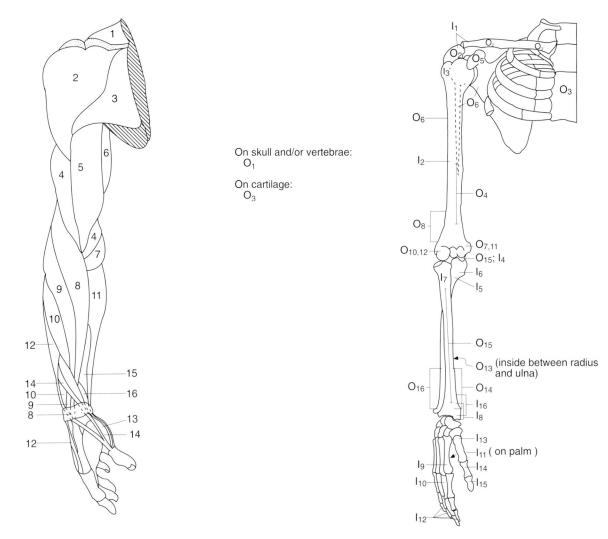

On skull and/or vertebrae:
  $O_1$

On cartilage:
  $O_3$

**Figure 55**  *Coloring book of arm, anterior (front). As appeared in* An Atlas of Anatomy for Artists. *By Fritz Schider. Dover Publications, Inc. NY.*

**Figure 56**  *Arm bones, anterior (front). As appeared in* An Atlas of Anatomy for Artists. *By Fritz Schider. Dover Publications, Inc. NY.*

| Name | Action | Origin | Insertion |
|------|--------|--------|-----------|
| 1. TRAPEZIUS | Brings scapulae together; moves them up and down; draws head backward | Occipital bone (on skull) down to thoracic vertebra #12 | Clavicle, acromion, & spine of scapula |
| 2. DELTOID | Moves arm forward and backward; moves arm to side away from body | Clavicle, acromion, & spine of scapula | Humerus |
| 3. PECTORALIS MAJOR | Moves arm forward and across body; rotates arm inward; lowers arm from vertical position | Clavicle, sternum, & spine of scapula | Humerus |
| 4. BRACHIALIS | Bends arm at elbow | Humerus | Ulna (inside elbow joint) |
| 5. BICEPS BRACHII | Bends arm at elbow | Coracoid process of scapula | Radius |
| 6. TRICEPS BRACHII | Straightens elbow | Humerus & scapula | Ulna (on olecranon) |
| 7. PRONATOR TERES | Pronates forearm | Humerus | Radius |
| 8. BRACHIORADIALIS | Bends arm at elbow; while arm is bent, pronates forearm; when arm is straight, supinates forearm | Humerus | Radius |
| 9. EXTENSOR CARPI RADIALIS LONGUS | Same as brachioradialis; plus straightens wrist and moves hand outward from side of body at wrist joint | Humerus | Metacarpal #2 |
| 10. EXTENSOR CARPI RADIALIS BREVIS | Straightens wrist; moves hand outward from side of body at wrist joint | Humerus | Metacarpal #3 |
| 11. FLEXOR CARPI RADIALIS | Bends wrist; moves hand toward side of body at wrist joint | Humerus | Metacarpals #2 & #3 |
| 12. EXTENSOR DIGITORUM | Straightens fingers & hand; spreads apart fingers | Humerus | By tendons to phalanges #2–5 |
| 13. ABDUCTOR POLLICIS LONGUS | Moves thumb out from side of hand; moves thumb forward | Radius & ulna | Metacarpal #1 |
| 14. EXTENSOR POLLICIS BREVIS | Straightens final two thumb joints | Radius | First phalanx of thumb |
| 15. FLEXOR POLLICIS LONGUS | Flexes thumb | Radius & ulna | Second phalanx of thumb |
| 16. PRONATOR QUADRATUS | Pronates hands | Lower ulna | Lower radius |

All actions and descriptive terms are defined with reference to the ANATOMICAL POSITION, that is, a figure standing erect, feet parallel and flat on the floor, eyes directed forward, arms at the sides, and palms facing forward.

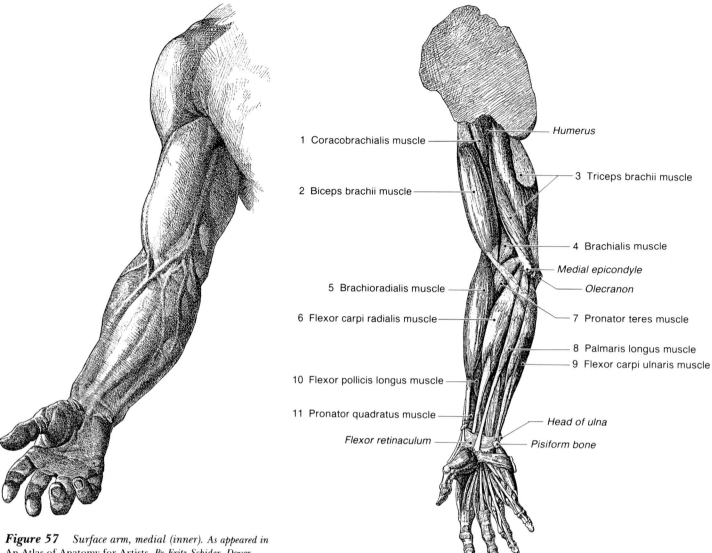

1 Coracobrachialis muscle — — *Humerus*

2 Biceps brachii muscle — — 3 Triceps brachii muscle

— 4 Brachialis muscle

*Medial epicondyle*

5 Brachioradialis muscle — *Olecranon*

6 Flexor carpi radialis muscle — 7 Pronator teres muscle

— 8 Palmaris longus muscle
— 9 Flexor carpi ulnaris muscle

10 Flexor pollicis longus muscle —

11 Pronator quadratus muscle — *Head of ulna*

*Flexor retinaculum* — *Pisiform bone*

**Figure 57** *Surface arm, medial (inner). As appeared in* An Atlas of Anatomy for Artists. *By Fritz Schider. Dover Publications, Inc. NY.*

**Figure 58** *Flayed arm, medial (inner). As appeared in* An Atlas of Anatomy for Artists. *By Fritz Schider. Dover Publications, Inc. NY.*

| Name | Action | Origin | Insertion |
|------|--------|--------|-----------|
| 1. CORACOBRACHIALIS | Raises arm forward; brings outstretched arm toward side of body | Scapula | Humerus |
| 2. BICEPS BRACHII | Bends arm at elbow | Coracoid process of scapula | Radius |
| 3. TRICEPS BRACHII | Straightens elbow | Humerus & scapula | Ulna (on olecranon) |
| 4. BRACHIALIS | Bends arm at elbow | Humerus | Ulna (inside elbow joint) |
| 5. BRACHIORADIALIS | Bends arm at elbow; while arm is bent, pronates forearm; when arm is straight, supinates forearm | Humerus | Radius |
| 6. FLEXOR CARPI RADIALIS | Bends wrist; moves hand toward side of body at wrist joint | Humerus | Metacarpals #2 & #3 |
| 7. PRONATOR TERES | Pronates forearm | Humerus | Radius |
| 8. PALMARIS LONGUS | Bends hand at wrist | Humerus | Palmar aponeurosis |
| 9. FLEXOR CARPI ULNARIS | Bends wrist; moves hand toward side of body at wrist joint | Humerus & olecranon | Pisiform bone and surrounding metacarpal bones |
| 10. FLEXOR POLLICIS LONGUS | Flexes thumb | Radius & ulna | Second phalanx of thumb |
| 11. PRONATOR QUADRATUS | Pronates hands | Lower ulna | Lower radius |

All actions and descriptive terms are defined with reference to the
ANATOMICAL POSITION, that is, a figure standing erect, feet
parallel and flat on the floor, eyes directed forward, arms at the sides,
and palms facing forward.

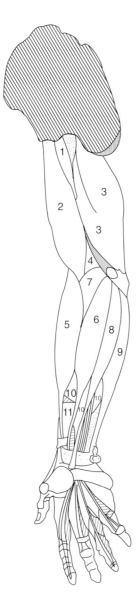

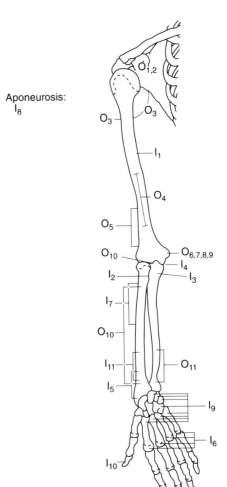

Aponeurosis:
$I_8$

*Figure 60*   *Albinus (1697–1770). Arms bones, medial (inner).*
[Albinus on Anatomy. *By Robert Beverly Hale and Terrance Coyle. Watson-Guptill Pub., NY. 1979. p. 28.]*

*Figure 59*   *Coloring book of arm, medial (inner). As appeared in* An Atlas of Anatomy for Artists. *By Fritz Schider. Dover Publications, Inc. NY.*

| Name | Action | Origin | Insertion |
|------|--------|--------|-----------|
| 1. CORACOBRACHIALIS | Raises arm forward; brings outstretched arm toward side of body | Scapula | Humerus |
| 2. BICEPS BRACHII | Bends arm at elbow | Coracoid process of scapula | Radius |
| 3. TRICEPS BRACHII | Straightens elbow | Humerus & scapula | Ulna (on olecranon) |
| 4. BRACHIALIS | Bends arm at elbow | Humerus | Ulna (inside elbow joint) |
| 5. BRACHIORADIALIS | Bends arm at elbow; while arm is bent, pronates forearm; when arm is straight, supinates forearm | Humerus | Radius |
| 6. FLEXOR CARPI RADIALIS | Bends wrist; moves hand toward side of body at wrist joint | Humerus | Metacarpals #2 & #3 |
| 7. PRONATOR TERES | Pronates forearm | Humerus | Radius |
| 8. PALMARIS LONGUS | Bends hand at wrist | Humerus | Palmar aponeurosis |
| 9. FLEXOR CARPI ULNARIS | Bends wrist; moves hand toward side of body at wrist joint | Humerus & olecranon | Pisiform bone and surrounding metacarpal bones |
| 10. FLEXOR POLLICIS LONGUS | Flexes thumb | Radius & ulna | Second phalanx of thumb |
| 11. PRONATOR QUADRATUS | Pronates hands | Lower ulna | Lower radius |

All actions and descriptive terms are defined with reference to the ANATOMICAL POSITION, that is, a figure standing erect, feet parallel and flat on the floor, eyes directed forward, arms at the sides, and palms facing forward.

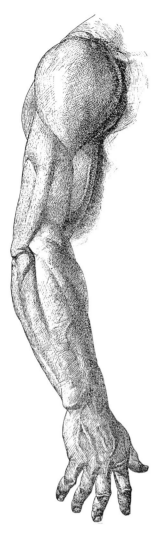

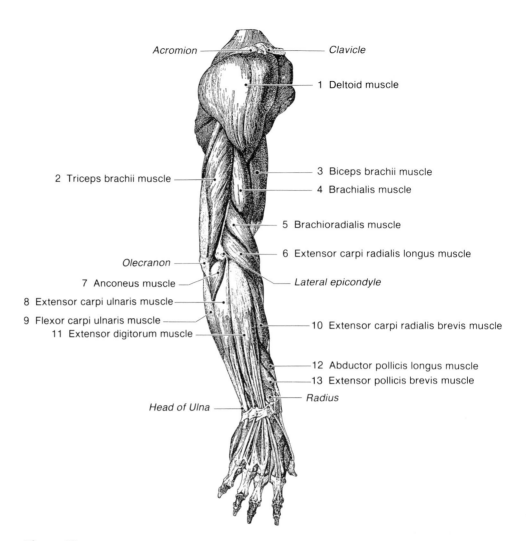

Acromion —
Clavicle

1 Deltoid muscle

2 Triceps brachii muscle —

3 Biceps brachii muscle
4 Brachialis muscle

5 Brachioradialis muscle

6 Extensor carpi radialis longus muscle

Olecranon —

7 Anconeus muscle —

Lateral epicondyle

8 Extensor carpi ulnaris muscle —

9 Flexor carpi ulnaris muscle —
11 Extensor digitorum muscle —

10 Extensor carpi radialis brevis muscle

12 Abductor pollicis longus muscle
13 Extensor pollicis brevis muscle

Head of Ulna —
Radius

***Figure 61*** *Surface arm, lateral (side). As appeared in* An Atlas of Anatomy for Artists. *By Fritz Schider. Dover Publications, Inc. NY.*

***Figure 62*** *Flayed arm, lateral (side). As appeared in* An Atlas of Anatomy for Artists. *By Fritz Schider. Dover Publications, Inc. NY.*

| Name | Action | Origin | Insertion |
|------|--------|--------|-----------|
| 1. DELTOID | Moves arm forward and backward; moves arm to side away from body | Clavicle, acromion, & spine of scapula | Humerus |
| 2. TRICEPS BRACHII | Straightens elbow | Humerus & scapula | Ulna (on olecranon) |
| 3. BICEPS BRACHII | Bends arm at elbow | Coracoid process of scapula | Radius |
| 4. BRACHIALIS | Bends arm at elbow | Humerus | Ulna (inside elbow joint) |
| 5. BRACHIORADIALIS | Bends arm at elbow; while arm is bent, pronates forearm; when arm is straight, supinates forearm | Humerus | Radius |
| 6. EXTENSOR CARPI RADIALIS LONGUS | Same as brachioradialis; plus straightens wrist and moves hand outward from side of body at wrist joint | Humerus | Metacarpal #2 |
| 7. ANCONEUS | Straightens elbow | Humerus | Ulna |
| 8. EXTENSOR CARPI ULNARIS | Straightens wrist; moves hand toward side of body at wrist joint | Humerus & ulna | Metacarpal #5 |
| 9. FLEXOR CARPI ULNARIS | Bends wrist; moves hand toward side of body at wrist joint | Humerus & olecranon | Pisiform bone & surrounding metacarpal bones |
| 10. EXTENSOR CARPI RADIALIS BREVIS | Straightens wrist; moves hand outward from side of body at wrist joint | Humerus | Metacarpal #3 |
| 11. EXTENSOR DIGITORUM | Straightens fingers & hand; spreads apart fingers | Humerus | By tendons to phalanges #2–5 |
| 12. ABDUCTOR POLLICIS LONGUS | Moves thumb out from side of hand; moves thumb forward | Radius & ulna | Metacarpal #1 |
| 13. EXTENSOR POLLICIS BREVIS | Straightens final two thumb joints | Radius | First phalanx of thumb |

All actions and descriptive terms are defined with reference to the
ANATOMICAL POSITION, that is, a figure standing erect, feet
parallel and flat on the floor, eyes directed forward, arms at the sides,
and palms facing forward.

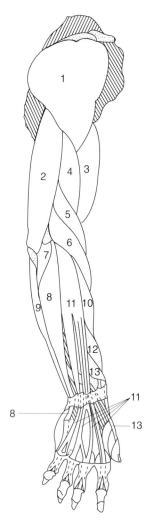

**Figure 63** *Coloring book of arm, lateral (side).* As appeared in An Atlas of Anatomy for Artists. *By Fritz Schider. Dover Publications, Inc. NY.*

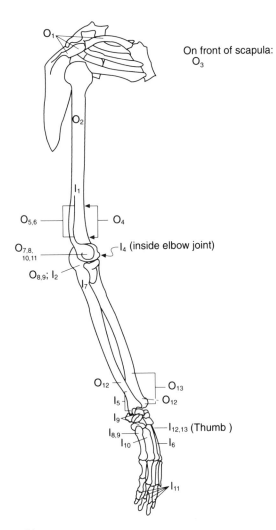

**Figure 64** *Arm bones, lateral (side).* As appeared in An Atlas of Anatomy for Artists. *By Fritz Schider. Dover Publications, Inc. NY.*

| Name | Action | Origin | Insertion |
|------|--------|--------|-----------|
| 1. DELTOID | Moves arm forward and backward; moves arm to side away from body | Clavicle, acromion, & spine of scapula | Humerus |
| 2. TRICEPS BRACHII | Straightens elbow | Humerus & scapula | Ulna (on olecranon) |
| 3. BICEPS BRACHII | Bends arm at elbow | Coracoid process of scapula | Radius |
| 4. BRACHIALIS | Bends arm at elbow | Humerus | Ulna (inside elbow joint) |
| 5. BRACHIORADIALIS | Bends arm at elbow; while arm is bent, pronates forearm; when arm is straight, supinates forearm | Humerus | Radius |
| 6. EXTENSOR CARPI RADIALIS LONGUS | Same as brachioradialis; plus straightens wrist and moves hand outward from side of body at wrist joint | Humerus | Metacarpal #2 |
| 7. ANCONEUS | Straightens elbow | Humerus | Ulna |
| 8. EXTENSOR CARPI ULNARIS | Straightens wrist; moves hand toward side of body at wrist joint | Humerus & ulna | Metacarpal #5 |
| 9. FLEXOR CARPI ULNARIS | Bends wrist; moves hand toward side of body at wrist joint | Humerus & olecranon | Pisiform bone & surrounding metacarpal bones |
| 10. EXTENSOR CARPI RADIALIS BREVIS | Straightens wrist; moves hand outward from side of body at wrist joint | Humerus | Metacarpal #3 |
| 11. EXTENSOR DIGITORUM | Straightens fingers & hand; spreads apart fingers | Humerus | By tendons to phalanges #2–5 |
| 12. ABDUCTOR POLLICIS LONGUS | Moves thumb out from side of hand; moves thumb forward | Radius & ulna | Metacarpal #1 |
| 13. EXTENSOR POLLICIS BREVIS | Straightens final two thumb joints | Radius | First phalanx of thumb |

All actions and descriptive terms are defined with reference to the ANATOMICAL POSITION, that is, a figure standing erect, feet parallel and flat on the floor, eyes directed forward, arms at the sides, and palms facing forward.

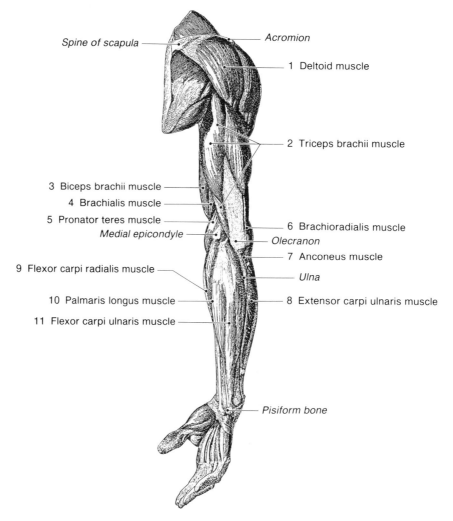

Spine of scapula
Acromion
1 Deltoid muscle
2 Triceps brachii muscle
3 Biceps brachii muscle
4 Brachialis muscle
5 Pronator teres muscle
Medial epicondyle
6 Brachioradialis muscle
Olecranon
7 Anconeus muscle
9 Flexor carpi radialis muscle
Ulna
10 Palmaris longus muscle
8 Extensor carpi ulnaris muscle
11 Flexor carpi ulnaris muscle
Pisiform bone

**Figure 65** *Surface arm, posterior (back). As appeared in* An Atlas of Anatomy for Artists. *By Fritz Schider. Dover Publications, Inc. NY.*

**Figure 66** *Flayed arm, posterior (back). As appeared in* An Atlas of Anatomy for Artists. *By Fritz Schider. Dover Publications, Inc. NY.*

| Name | Action | Origin | Insertion |
|------|--------|--------|-----------|
| 1. DELTOID | Moves arm forward and backward; moves arm to side away from body | Clavicle, acromion, & spine of scapula | Humerus |
| 2. TRICEPS BRACHII | Straightens elbow | Humerus & scapula | Ulna (on olecranon) |
| 3. BICEPS BRACHII | Bends arm at elbow | Coracoid process of scapula | Radius |
| 4. BRACHIALIS | Bends arm at elbow | Humerus | Ulna (inside elbow joint) |
| 5. PRONATOR TERES | Pronates forearm | Humerus | Radius |
| 6. BRACHIORADIALIS | Bends arm at elbow; while arm is bent, pronates forearm; when arm is straight, supinates forearm | Humerus | Radius |
| 7. ANCONEUS | Straightens elbow | Humerus | Ulna |
| 8. EXTENSOR CARPI ULNARIS | Straightens wrist; moves hand toward side of body at wrist | Humerus & ulna | Metacarpal #5 |
| 9. FLEXOR CARPI RADIALIS | Bends wrist; moves hand toward side of body at wrist joint | Humerus | Metacarpals #2 & #3 |
| 10. PALMARIS LONGUS | Flexes hand at wrist | Humerus | Palmar aponeurosis |
| 11. FLEXOR CARPI ULNARIS | Bends wrist; moves hand toward side of body at wrist joint | Humerus & olecranon | Pisiform bone and surrounding metacarpal bones |

All actions and descriptive terms are defined with reference to the ANATOMICAL POSITION, that is, a figure standing erect, feet parallel and flat on the floor, eyes directed forward, arms at the sides, and palms facing forward.

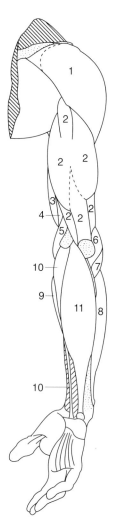

**Figure 67** *Coloring book of arm, posterior (back). As appeared in* An Atlas of Anatomy for Artists. *By Fritz Schider. Dover Publications, Inc. NY.*

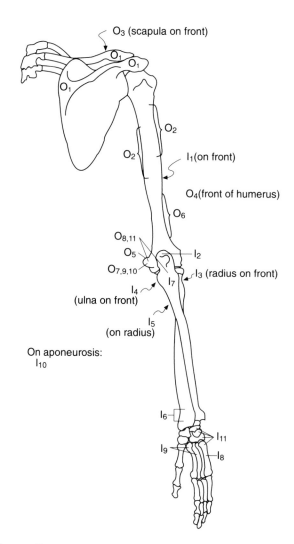

$O_3$ (scapula on front)

$O_1$  $O_1$

$O_1$

$O_2$

$O_2$

$I_1$(on front)

$O_4$(front of humerus)

$O_6$

$O_{8,11}$

$O_5$

$I_2$

$O_{7,9,10}$

$I_3$ (radius on front)

$I_7$

$I_4$

(ulna on front)

$I_5$
(on radius)

On aponeurosis:
$I_{10}$

$I_6$

$I_{11}$

$I_9$

$I_8$

**Figure 68** *Arm bones, posterior (back). As appeared in* An Atlas of Anatomy for Artists. *By Fritz Schider. Dover Publications, Inc. NY.*

| Name | Action | Origin | Insertion |
|------|--------|--------|-----------|
| 1. DELTOID | Moves arm forward and backward; moves arm to side away from body | Clavicle, acromion, & spine of scapula | Humerus |
| 2. TRICEPS BRACHII | Straightens elbow | Humerus & scapula | Ulna (on olecranon) |
| 3. BICEPS BRACHII | Bends arm at elbow | Coracoid process of scapula | Radius |
| 4. BRACHIALIS | Bends arm at elbow | Humerus | Ulna (inside elbow joint) |
| 5. PRONATOR TERES | Pronates forearm | Humerus | Radius |
| 6. BRACHIORADIALIS | Bends arm at elbow; while arm is bent, pronates forearm; when arm is straight, supinates forearm | Humerus | Radius |
| 7. ANCONEUS | Straightens elbow | Humerus | Ulna |
| 8. EXTENSOR CARPI ULNARIS | Straightens wrist; moves hand toward side of body at wrist | Humerus & ulna | Metacarpal #5 |
| 9. FLEXOR CARPI RADIALIS | Bends wrist; moves hand toward side of body at wrist joint | Humerus | Metacarpals #2 & #3 |
| 10. PALMARIS LONGUS | Flexes hand at wrist | Humerus | Palmar aponeurosis |
| 11. FLEXOR CARPI ULNARIS | Bends wrist; moves hand toward side of body at wrist joint | Humerus & olecranon | Pisiform bone and surrounding metacarpal bones |

All actions and descriptive terms are defined with reference to the ANATOMICAL POSITION, that is, a figure standing erect, feet parallel and flat on the floor, eyes directed forward, arms at the sides, and palms facing forward.

## Arms

Working with the arms/hands and legs/feet is much less complicated than with the head and torso. Choose an arm of the skeleton, front, side or back. Draw the *sensitizing* blind contour, again shielding the paper on which you draw the image. Next, the *structure* drawing, using the sighting, gesture, and volumetric shapes. Last, the *superimposition* of bones into arms. Use the book for referencing bulges and fleshy indentations. You will again find discrepancies between the length and width of the bones you just drew and with the model who is posing. Adjust the proportions accordingly, trying to be as accurate as possible in placing the bones into the fleshy muscle forms of the body. Remember to use your own body to find joints or to feel where in the arm the bones rest between or among the muscles. Sometimes bones are near the surface, other places the bones are imbedded. Your own body is certainly another resource available to you.                                      ▲

***Figure 69***   *Blind contour of an arm.*

**Figure 71**   *Superimposition of bones into the arm.*

**Figure 70**   *Structure and sighting of an arm.*

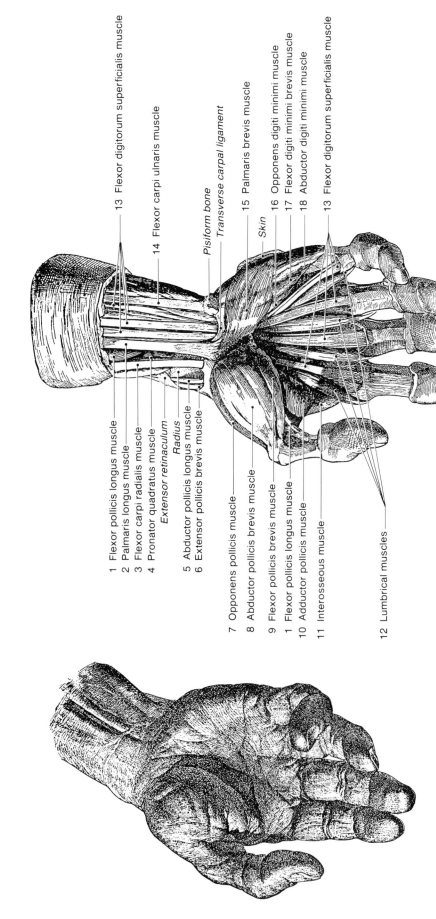

13 Flexor digitorum superficialis muscle

14 Flexor carpi ulnaris muscle

Pisiform bone
Transverse carpal ligament

Skin

15 Palmaris brevis muscle
16 Opponens digiti minimi muscle
17 Flexor digiti minimi brevis muscle
18 Abductor digiti minimi muscle
13 Flexor digitorum superficialis muscle

1 Flexor pollicis longus muscle
2 Palmaris longus muscle
3 Flexor carpi radialis muscle
4 Pronator quadratus muscle
Extensor retinaculum
Radius
5 Abductor pollicis longus muscle
6 Extensor pollicis brevis muscle

7 Opponens pollicis muscle
8 Abductor pollicis brevis muscle
9 Flexor pollicis brevis muscle
1 Flexor pollicis longus muscle
10 Adductor pollicis muscle
11 Interosseous muscle

12 Lumbrical muscles

**Figure 73** *Flayed right hand, anterior (front). As appeared in An Atlas of Anatomy for Artists. By Fritz Schider. Dover Publications, Inc. NY.*

**Figure 72** *Surface hand, anterior (front). As appeared in An Atlas of Anatomy for Artists. By Fritz Schider. Dover Publications, Inc. NY.*

| Name | Action | Origin | Insertion |
|---|---|---|---|
| 1. FLEXOR POLLICIS LONGUS | Flexes thumb | Radius & ulna | Second phalanx of the thumb |
| 2. PALMARIS LONGUS | Flexes hand at wrist | Humerus | Palmar aponeurosis |
| 3. FLEXOR CARPI RADIALIS | Bends wrist; moves hand toward side of body at wrist joint | Humerus | Metacarpals #2 & #3 |
| 4. PRONATOR QUADRATUS | Pronates hand | Ulna (at wrist) | Radius (at wrist) |
| 5. ABDUCTOR POLLICIS LONGUS | Moves thumb away from side of hand; moves thumb forward | Radius & ulna | Metacarpal #1 |
| 6. EXTENSOR POLLICIS BREVIS | Straightens final two thumb joints | Radius | First phalanx of thumb |
| 7. OPPONENS POLLICIS | Moves thumb toward side of hand; moves thumb across palm to meet little finger | Carpal ligament and trapezium (a carpal bone) | Metacarpal #1 |
| 8. ABDUCTOR POLLICIS BREVIS | Moves thumb away from side of hand; moves thumb forward | Carpal ligament and trapezium (a carpal bone) | First phalanx of the thumb |
| 9. FLEXOR POLLICIS BREVIS | Moves thumb forward; bends thumb at first joint; straightens second joint | Scaphoid of the novicular bone | Metacarpal #1 (base of thumb) |
| 10. ADDUCTOR POLLICIS | Moves thumb toward side of hand | Second and third metacarpals; capitate (a carpal bone) | First phalanx of thumb |
| 11. INTEROSSEI | Spread apart and close fingers; flexes index finger | Adjacent sides of metacarpals #2, 4, & 5 | First and second phalanges of the fingers and extensor digitorum |
| 12. LUMBRICALES | Bend and straighten fingers | Tendons of flexor digitorum profundus | Digits of the fingers |
| 13. FLEXOR DIGITORUM SUPERFICIALIS | Helps bend each finger, joint by joint | Humerus, ulna, & radius | Second phalanges of #2–5 |
| 14. FLEXOR CARPI ULNARIS | Bends wrist; moves hand toward side of body at wrist joint | Humerus & olecranon | Pisiform bone and surrounding metacarpal bones |
| 15. PALMARIS BREVIS | Corrugates skin of palm to improve grip | Palmar aponeurosis | Skin of palm |
| 16. OPPONENS DIGITI MINIMI | Helps move little finger across palm to meet thumb; also moves little finger when palm is cupped | Flexor retinaculum and hamate (a carpal bone) | Metacarpal #5 |
| 17. FLEXOR DIGITI MINIMI BREVIS | Bends little finger | Flexor retinaculum and hamate (a carpal bone) | First phalanx of little finger |
| 18. ABDUCTOR DIGITI MINIMI | Moves little finger out from other fingers | Pisiform bone | First phalanx of little finger |

All actions and descriptive terms are defined with reference to the ANATOMICAL POSITION, that is, a figure standing erect, feet parallel and flat on the floor, eyes directed forward, arms at the sides, and palms facing forward.

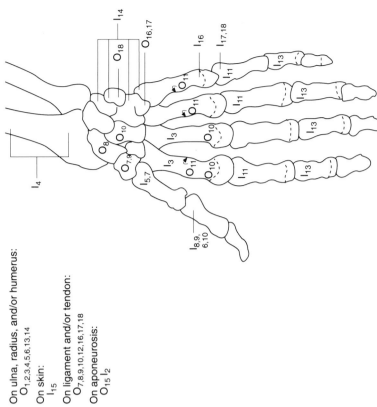

On ulna, radius, and/or humerus:
O$_{1,2,3,4,5,6,13,14}$

On skin:
I$_{15}$

On ligament and/or tendon:
O$_{7,8,9,10,12,16,17,18}$

On aponeurosis:
O$_{15}$ I$_2$

**Figure 75** *Albinus (1697–1770). Right hand bones, anterior (front). [Albinus on Anatomy. By Robert Beverly Hale and Terrance Coyle. Watson-Guptill Pub., NY. 1979. p. 28.]*

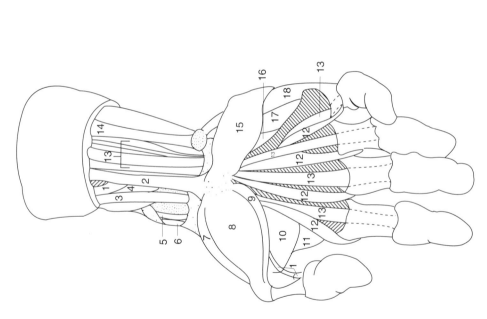

**Figure 74** *Coloring book of hand, anterior (front). As appeared in An Atlas of Anatomy for Artists. By Fritz Schider. Dover Publications, Inc. NY.*

| Name | Action | Origin | Insertion |
|------|--------|--------|-----------|
| 1. FLEXOR POLLICIS LONGUS | Flexes thumb | Radius & ulna | Second phalanx of the thumb |
| 2. PALMARIS LONGUS | Flexes hand at wrist | Humerus | Palmar aponeurosis |
| 3. FLEXOR CARPI RADIALIS | Bends wrist; moves hand toward side of body at wrist joint | Humerus | Metacarpals #2 & #3 |
| 4. PRONATOR QUADRATUS | Pronates hand | Ulna (at wrist) | Radius (at wrist) |
| 5. ABDUCTOR POLLICIS LONGUS | Moves thumb away from side of hand; moves thumb forward | Radius & ulna | Metacarpal #1 |
| 6. EXTENSOR POLLICIS BREVIS | Straightens final two thumb joints | Radius | First phalanx of thumb |
| 7. OPPONENS POLLICIS | Moves thumb toward side of hand; moves thumb across palm to meet little finger | Carpal ligament and trapezium (a carpal bone) | Metacarpal #1 |
| 8. ABDUCTOR POLLICIS BREVIS | Moves thumb away from side of hand; moves thumb forward | Carpal ligament and trapezium (a carpal bone) | First phalanx of the thumb |
| 9. FLEXOR POLLICIS BREVIS | Moves thumb forward; bends thumb at first joint; straightens second joint | Scaphoid of the novicular bone | Metacarpal #1 (base of thumb) |
| 10. ADDUCTOR POLLICIS | Moves thumb toward side of hand | Second and third metacarpals; capitate (a carpal bone) | First phalanx of thumb |
| 11. INTEROSSEI | Spread apart and close fingers; flexes index finger | Adjacent sides of metacarpals #2, 4, & 5 | First and second phalanges of the fingers and extensor digitorum |
| 12. LUMBRICALES | Bend and straighten fingers | Tendons of flexor digitorum profundus | Digits of the fingers |
| 13. FLEXOR DIGITORUM SUPERFICIALIS | Helps bend each finger, joint by joint | Humerus, ulna, & radius | Second phalanges of #2–5 |
| 14. FLEXOR CARPI ULNARIS | Bends wrist; moves hand toward side of body at wrist joint | Humerus & olecranon | Pisiform bone and surrounding metacarpal bones |
| 15. PALMARIS BREVIS | Corrugates skin of palm to improve grip | Palmar aponeurosis | Skin of palm |
| 16. OPPONENS DIGITI MINIMI | Helps move little finger across palm to meet thumb; also moves little finger when palm is cupped | Flexor retinaculum and hamate (a carpal bone) | Metacarpal #5 |
| 17. FLEXOR DIGITI MINIMI BREVIS | Bends little finger | Flexor retinaculum and hamate (a carpal bone) | First phalanx of little finger |
| 18. ABDUCTOR DIGITI MINIMI | Moves little finger out from other fingers | Pisiform bone | First phalanx of little finger |

All actions and descriptive terms are defined with reference to the ANATOMICAL POSITION, that is, a figure standing erect, feet parallel and flat on the floor, eyes directed forward, arms at the sides, and palms facing forward.

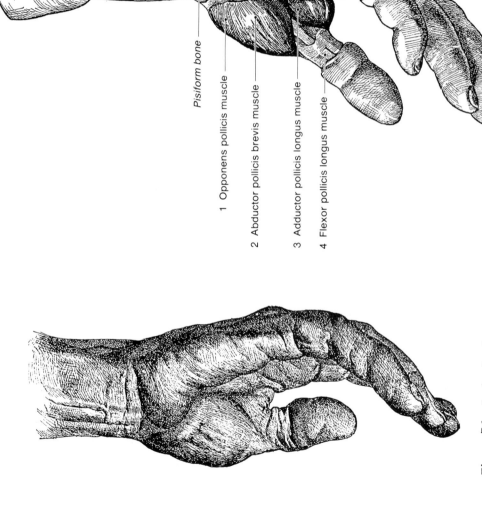

5 Flexor carpi ulnaris muscle
6 Extensor carpi ulnaris muscle
7 Extensor digitorum muscle

*Head of Ulna*

*Extensor retinaculum*

6 Extensor carpi ulnaris muscle
8 Palmaris brevis muscle
9 Abductor digiti minimi muscle
10 Extensor digiti minimi muscle
7 Extensor digitorum muscle

*Metacarpal head*

*Pisiform bone*

1 Opponens pollicis muscle
2 Abductor pollicis brevis muscle
3 Adductor pollicis longus muscle
4 Flexor pollicis longus muscle

**Figure 77**  *Flayed hand, medial (inner). As appeared in An Atlas of Anatomy for Artists. By Fritz Schider. Dover Publications, Inc. NY.*

**Figure 76**  *Surface hand, medial (inner). As appeared in An Atlas of Anatomy for Artists. By Fritz Schider. Dover Publications, Inc. NY.*

| Name | Action | Origin | Insertion |
|---|---|---|---|
| 1. OPPONENS POLLICIS | Moves thumb toward side of hand; moves thumb across palm to meet little finger | Carpal ligament and trapezium (a carpal bone) | Metacarpal #1 |
| 2. ABDUCTOR POLLICIS BREVIS | Moves thumb away from side of hand; moves thumb forward | Carpal ligament and trapezium (a tubercle of carpal bone) | First phalanx of thumb |
| 3. ADDUCTOR POLLICIS LONGUS | Moves thumb away from side of hand; moves thumb forward | Radius & ulna | Metacarpal #1 |
| 4. FLEXOR POLLICIS LONGUS | Flexes thumb | Radius & ulna | Second phalanx of the thumb |
| 5. FLEXOR CARPI ULNARIS | Bends wrist; moves hand toward side of body at wrist joint | Humerus & olecranon | Pisiform bone and surrounding metacarpal bones |
| 6. EXTENSOR CARPI ULNARIS | Bends wrist; moves hand toward side of body at wrist joint | Humerus & olecranon | Pisiform bone and surrounding metacarpal bones |
| 7. EXTENSOR DIGITORUM | Straightens fingers and hand; spreads apart fingers | Humerus | By tendons to phalanges #2–5 |
| 8. PALMARIS BREVIS | Corrugates skin of palm to improve grip | Palmar aponeurosis | Skin of palm |
| 9. ABDUCTOR DIGITI MINIMI | Moves little finger out from other fingers | Pisiform bone | First phalanx of little finger |
| 10. EXTENSOR DIGITI MINIMI | Straightens little finger | Humerus | First phalanx of the little finger |

All actions and descriptive terms are defined with reference to the ANATOMICAL POSITION, that is, a figure standing erect, feet parallel and flat on the floor, eyes directed forward, arms at the sides, and palms facing forward.

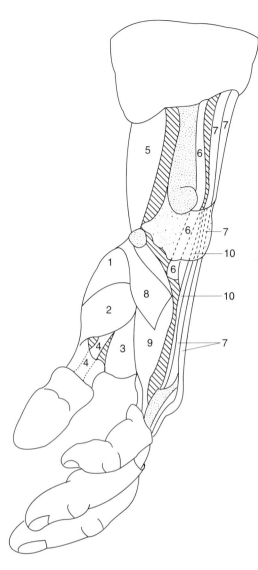

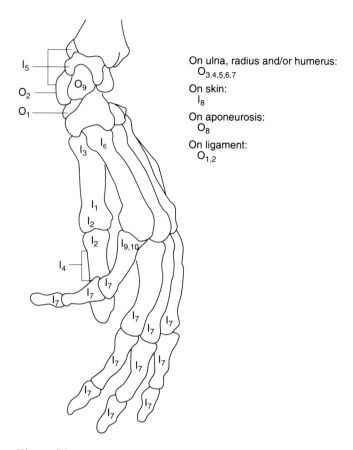

On ulna, radius and/or humerus:
$O_{3,4,5,6,7}$

On skin:
$I_8$

On aponeurosis:
$O_8$

On ligament:
$O_{1,2}$

**Figure 79**  *Albinus (1697–1770). Hand bones, medial (inner).*
[Albinus on Anatomy. *By Robert Beverly Hale and Terrance Coyle. Watson-Guptill Pub., NY. 1979. p. 32.]*

**Figure 78**  *Coloring book of hand, medial (inner). As appeared in*
An Atlas of Anatomy for Artists. *By Fritz Schider. Dover Publications, Inc. NY.*

| Name | Action | Origin | Insertion |
|------|--------|--------|-----------|
| 1. OPPONENS POLLICIS | Moves thumb toward side of hand; moves thumb across palm to meet little finger | Carpal ligament and trapezium (a carpal bone) | Metacarpal #1 |
| 2. ABDUCTOR POLLICIS BREVIS | Moves thumb away from side of hand; moves thumb forward | Carpal ligament and trapezium (a tubercle of carpal bone) | First phalanx of thumb |
| 3. ADDUCTOR POLLICIS LONGUS | Moves thumb away from side of hand; moves thumb forward | Radius & ulna | Metacarpal #1 |
| 4. FLEXOR POLLICIS LONGUS | Flexes thumb | Radius & ulna | Second phalanx of the thumb |
| 5. FLEXOR CARPI ULNARIS | Bends wrist; moves hand toward side of body at wrist joint | Humerus & olecranon | Pisiform bone and surrounding metacarpal bones |
| 6. EXTENSOR CARPI ULNARIS | Bends wrist; moves hand toward side of body at wrist joint | Humerus & olecranon | Pisiform bone and surrounding metacarpal bones |
| 7. EXTENSOR DIGITORUM | Straightens fingers and hand; spreads apart fingers | Humerus | By tendons to phalanges #2–5 |
| 8. PALMARIS BREVIS | Corrugates skin of palm to improve grip | Palmar aponeurosis | Skin of palm |
| 9. ABDUCTOR DIGITI MINIMI | Moves little finger out from other fingers | Pisiform bone | First phalanx of little finger |
| 10. EXTENSOR DIGITI MINIMI | Straightens little finger | Humerus | First phalanx of the little finger |

All actions and descriptive terms are defined with reference to the ANATOMICAL POSITION, that is, a figure standing erect, feet parallel and flat on the floor, eyes directed forward, arms at the sides, and palms facing forward.

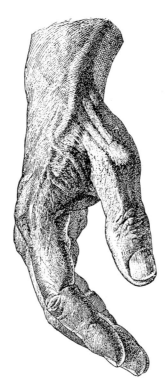

**Figure 80**   *Surface hand, lateral (side). As appeared in* An Atlas of Anatomy for Artists. *By Fritz Schider. Dover Publications, Inc. NY.*

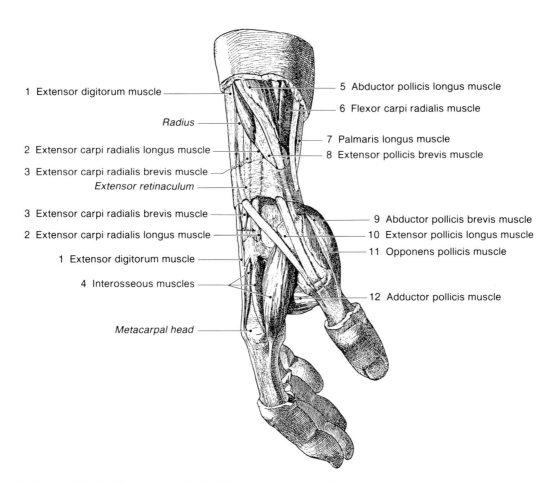

1 Extensor digitorum muscle

*Radius*

2 Extensor carpi radialis longus muscle

3 Extensor carpi radialis brevis muscle

*Extensor retinaculum*

3 Extensor carpi radialis brevis muscle

2 Extensor carpi radialis longus muscle

1 Extensor digitorum muscle

4 Interosseous muscles

*Metacarpal head*

5 Abductor pollicis longus muscle

6 Flexor carpi radialis muscle

7 Palmaris longus muscle

8 Extensor pollicis brevis muscle

9 Abductor pollicis brevis muscle

10 Extensor pollicis longus muscle

11 Opponens pollicis muscle

12 Adductor pollicis muscle

**Figure 81**   *Flayed hand, lateral (side). As appeared in* An Atlas of Anatomy for Artists. *By Fritz Schider. Dover Publications, Inc. NY.*

| Name | Action | Origin | Insertion |
|---|---|---|---|
| 1. EXTENSOR DIGITORUM | Straightens fingers and hand; spreads apart fingers | Humerus | By tendons to phalanges #2–5 |
| 2. EXTENSOR CARPI RADIALIS LONGUS | Same as brachioradialis; straightens wrist and moves hand away from side of body at wrist joint | Humerus | Metacarpal #2 |
| 3. EXTENSOR CARPI RADIALIS BREVIS | Straightens wrist; moves hand outward from side of body at wrist joint | Humerus | Metacarpal #3 |
| 4. INTEROSSEI | Spread apart and close fingers; flexes index finger | Adjacent sides of metacarpals #2, 4, & 5 | First and second phalanx of the fingers and extensor digitorum |
| 5. ABDUCTOR POLLICIS LONGUS | Moves thumb away from side of hand; moves thumb forward | Radius & ulna | Metacarpal #1 |
| 6. FLEXOR CARPI RADIALIS | Bends wrist; moves hand toward side of body at wrist joint | Humerus | Metacarpals #2 & #3 |
| 7. PALMARIS LONGUS | Flexes hand at wrist | Humerus | Palmar aponeurosis |
| 8. EXTENSOR POLLICIS BREVIS | Straightens final two thumb joints | Radius | First phalanx of thumb |
| 9. ABDUCTOR POLLICIS BREVIS | Moves thumb away from side of hand; moves thumb forward | Carpal ligament and trapezium (a carpal bone) | First phalanx of thumb |
| 10. EXTENSOR POLLICIS LONGUS | Straightens thumb; moves hand toward body at wrist joint | Radius | Second phalanx of thumb |
| 11. OPPONENS POLLICIS | Moves thumb toward side of hand; moves thumb across palm to meet little finger | Carpal ligament and trapezium (a carpal bone) | Metacarpal #1 |
| 12. ADDUCTOR POLLICIS | Moves thumb toward side of hand | Second and third metacarpals; capitate (a carpal bone) | First phalanx of the thumb |

All actions and descriptive terms are defined with reference to the ANATOMICAL POSITION, that is, a figure standing erect, feet parallel and flat on the floor, eyes directed forward, arms at the sides, and palms facing forward.

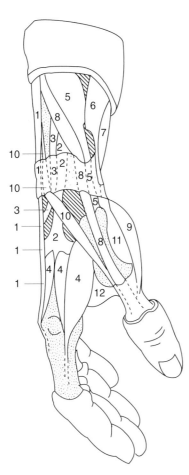

**Figure 82** *Coloring book of hand, lateral (side). As appeared in* An Atlas of Anatomy for Artists. *By Fritz Schider. Dover Publications, Inc. NY.*

On humerus, ulna, and/or radius:
$O_{1,2,3,5,6,7,8,10}$

On ligaments and tendons:
$O_{9,11}; I_4$

On aponeurosis:
$I_7$

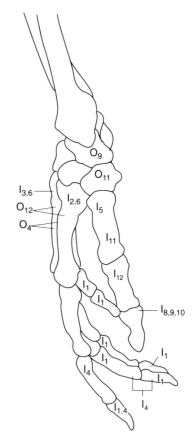

**Figure 83** *Hand bones, lateral (side). As appeared in* An Atlas of Anatomy for Artists. *By Fritz Schider. Dover Publications, Inc. NY.*

| Name | Action | Origin | Insertion |
|------|--------|--------|-----------|
| 1. EXTENSOR DIGITORUM | Straightens fingers and hand; spreads apart fingers | Humerus | By tendons to phalanges #2–5 |
| 2. EXTENSOR CARPI RADIALIS LONGUS | Same as brachioradialis; straightens wrist and moves hand away from side of body at wrist joint | Humerus | Metacarpal #2 |
| 3. EXTENSOR CARPI RADIALIS BREVIS | Straightens wrist; moves hand outward from side of body at wrist joint | Humerus | Metacarpal #3 |
| 4. INTEROSSEI | Spread apart and close fingers; flexes index finger | Adjacent sides of metacarpals #2, 4, & 5 | First and second phalanx of the fingers and extensor digitorum |
| 5. ABDUCTOR POLLICIS LONGUS | Moves thumb away from side of hand; moves thumb forward | Radius & ulna | Metacarpal #1 |
| 6. FLEXOR CARPI RADIALIS | Bends wrist; moves hand toward side of body at wrist joint | Humerus | Metacarpals #2 & #3 |
| 7. PALMARIS LONGUS | Flexes hand at wrist | Humerus | Palmar aponeurosis |
| 8. EXTENSOR POLLICIS BREVIS | Straightens final two thumb joints | Radius | First phalanx of thumb |
| 9. ABDUCTOR POLLICIS BREVIS | Moves thumb away from side of hand; moves thumb forward | Carpal ligament and trapezium (a carpal bone) | First phalanx of thumb |
| 10. EXTENSOR POLLICIS LONGUS | Straightens thumb; moves hand toward body at wrist joint | Radius | Second phalanx of thumb |
| 11. OPPONENS POLLICIS | Moves thumb toward side of hand; moves thumb across palm to meet little finger | Carpal ligament and trapezium (a carpal bone) | Metacarpal #1 |
| 12. ADDUCTOR POLLICIS | Moves thumb toward side of hand | Second and third metacarpals; capitate (a carpal bone) | First phalanx of the thumb |

All actions and descriptive terms are defined with reference to the ANATOMICAL POSITION, that is, a figure standing erect, feet parallel and flat on the floor, eyes directed forward, arms at the sides, and palms facing forward.

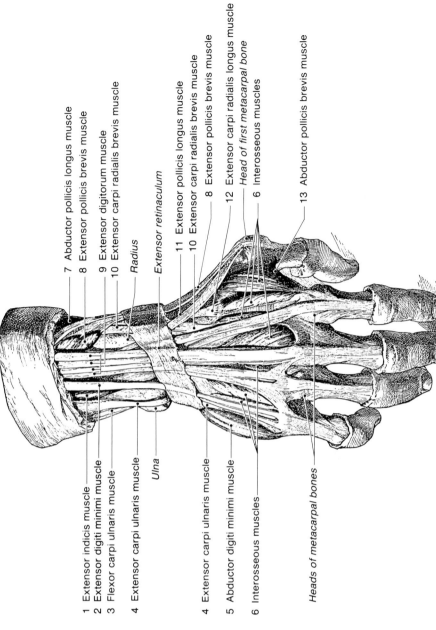

7 Abductor pollicis longus muscle
8 Extensor pollicis brevis muscle
9 Extensor digitorum muscle
10 Extensor carpi radialis brevis muscle

Radius

Extensor retinaculum

11 Extensor pollicis longus muscle
10 Extensor carpi radialis brevis muscle
8 Extensor pollicis brevis muscle
12 Extensor carpi radialis longus muscle
Head of first metacarpal bone
6 Interosseous muscles

13 Abductor pollicis brevis muscle

1 Extensor indicis muscle
2 Extensor digiti minimi muscle
3 Flexor carpi ulnaris muscle

4 Extensor carpi ulnaris muscle

Ulna

4 Extensor carpi ulnaris muscle

5 Abductor digiti minimi muscle

6 Interosseous muscles

Heads of metacarpal bones

**Figure 85** *Flayed hand, posterior (back). As appeared in* An Atlas of Anatomy for Artists. *By Fritz Schider. Dover Publications, Inc. NY.*

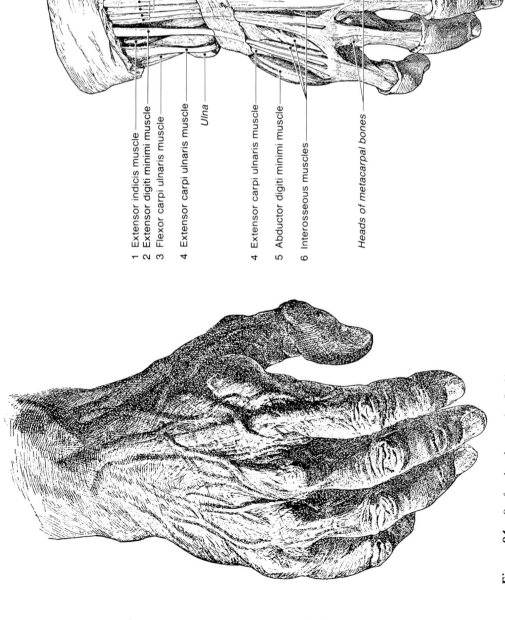

**Figure 84** *Surface hand, posterior (back). As appeared in* An Atlas of Anatomy for Artists. *By Fritz Schider. Dover Publications, Inc. NY.*

| Name | Action | Origin | Insertion |
|---|---|---|---|
| 1. EXTENSOR INDICIS | Straightens index finger | Ulna | First phalanx of index finger |
| 2. EXTENSOR DIGITI MINIMI | Straightens little finger | Humerus | First phalanx of the little finger |
| 3. FLEXOR CARPI ULNARIS | Bends wrist; moves hand toward side of body at wrist joint | Humerus & olecranon | Pisiform bone and surrounding metacarpal bones |
| 4. EXTENSOR CARPI ULNARIS | Bends wrist; moves hand toward side of body at wrist joint | Humerus & olecranon | Pisiform bone and surrounding metacarpal bones |
| 5. ABDUCTOR DIGITI MINIMI | Moves little finger out from other fingers | Pisiform bone | First phalanx of little finger |
| 6. INTEROSSEI | Spread apart and close fingers; flexes index finger | Adjacent sides of metacarpals | First and second phalanges of the fingers and extensor digitorum |
| 7. ABDUCTOR POLLICIS LONGUS | Moves thumb away from side of hand; moves thumb forward | Radius & ulna | Metacarpal #1 |
| 8. EXTENSOR POLLICIS BREVIS | Straightens final two thumb joints | Radius | First phalanx of the thumb |
| 9. EXTENSOR DIGITORUM | Straightens fingers and hand; spreads apart fingers | Humerus | By tendons to phalanges #2–5 |
| 10. EXTENSOR CARPI RADIALIS BREVIS | Straightens wrist; moves hand outward from side of body at wrist joint | Humerus | Metacarpal #3 |
| 11. EXTENSOR POLLICIS LONGUS | Straightens thumb; moves hand toward body at wrist joint | Radius | Second phalanx of thumb |
| 12. EXTENSOR CARPI RADIALIS LONGUS | Same as brachioradialis; straightens wrist and moves hand away from side of body at wrist joint | Humerus | Metacarpal #2 |
| 13. ABDUCTOR POLLICIS BREVIS | Moves thumb away from side of hand; moves thumb forward | Carpal ligament and trapezium (a scaphoid of carpal bone) | First phalanx of thumb |

All actions and descriptive terms are defined with reference to the ANATOMICAL POSITION, that is, a figure standing erect, feet parallel and flat on the floor, eyes directed forward, arms at the sides, and palms facing forward.

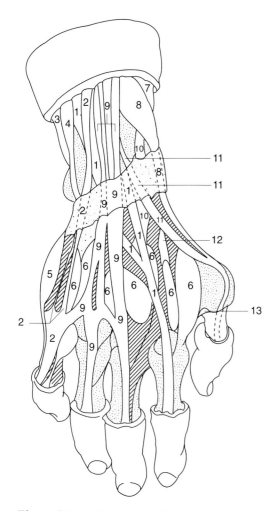

**Figure 86** *Coloring book of hand, posterior (back). As appeared in* An Atlas of Anatomy for Artists. *By Fritz Schider. Dover Publications, Inc. NY.*

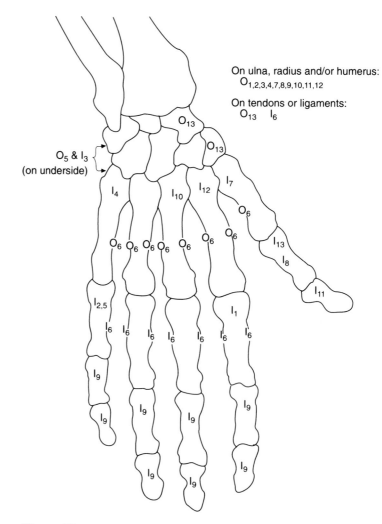

On ulna, radius and/or humerus:
$O_{1,2,3,4,7,8,9,10,11,12}$

On tendons or ligaments:
$O_{13}$    $I_6$

**Figure 87** *Albinus (1697–1770). Hand bones, posterior (back). [Albinus on Anatomy. By Robert Beverly Hale and Terrance Coyle. Watson-Guptill Pub., NY. 1979. p. 30.]*

| Name | Action | Origin | Insertion |
|---|---|---|---|
| 1. EXTENSOR INDICIS | Straightens index finger | Ulna | First phalanx of index finger |
| 2. EXTENSOR DIGITI MINIMI | Straightens little finger | Humerus | First phalanx of the little finger |
| 3. FLEXOR CARPI ULNARIS | Bends wrist; moves hand toward side of body at wrist joint | Humerus & olecranon | Pisiform bone and surrounding metacarpal bones |
| 4. EXTENSOR CARPI ULNARIS | Bends wrist; moves hand toward side of body at wrist joint | Humerus & olecranon | Pisiform bone and surrounding metacarpal bones |
| 5. ABDUCTOR DIGITI MINIMI | Moves little finger out from other fingers | Pisiform bone | First phalanx of little finger |
| 6. INTEROSSEI | Spread apart and close fingers; flexes index finger | Adjacent sides of metacarpals | First and second phalanges of the fingers and extensor digitorum |
| 7. ABDUCTOR POLLICIS LONGUS | Moves thumb away from side of hand; moves thumb forward | Radius & ulna | Metacarpal #1 |
| 8. EXTENSOR POLLICIS BREVIS | Straightens final two thumb joints | Radius | First phalanx of the thumb |
| 9. EXTENSOR DIGITORUM | Straightens fingers and hand; spreads apart fingers | Humerus | By tendons to phalanges #2–5 |
| 10. EXTENSOR CARPI RADIALIS BREVIS | Straightens wrist; moves hand outward from side of body at wrist joint | Humerus | Metacarpal #3 |
| 11. EXTENSOR POLLICIS LONGUS | Straightens thumb; moves hand toward body at wrist joint | Radius | Second phalanx of thumb |
| 12. EXTENSOR CARPI RADIALIS LONGUS | Same as brachioradialis; straightens wrist and moves hand away from side of body at wrist joint | Humerus | Metacarpal #2 |
| 13. ABDUCTOR POLLICIS BREVIS | Moves thumb away from side of hand; moves thumb forward | Carpal ligament and trapezium (a scaphoid of carpal bone) | First phalanx of thumb |

All actions and descriptive terms are defined with reference to the
ANATOMICAL POSITION, that is, a figure standing erect, feet
parallel and flat on the floor, eyes directed forward, arms at the sides,
and palms facing forward.

## Hands

Use your own hand as a model for this drawing. You know it best, have it with you all the time, and can poke and feel it as needed for these exercises. As you move through these exercises, remember that you give yourself information. Your concentrated efforts will pay off. Learning to play a musical instrument is satisfactory in steps. Once you master both the instrument and music, you can play professionally, improvise, or create music in the privacy of your home. So, too, with drawing. But there are many steps through which to move. You can do it, but you need to do it often. So, begin with the blind contour, move to the gesture for sighting and volume. Finally the bones are again imposed into the flesh. ▲

*Figure 88* *Blind contour of a hand.*

**Figure 89**   *Structural drawing of a hand.*

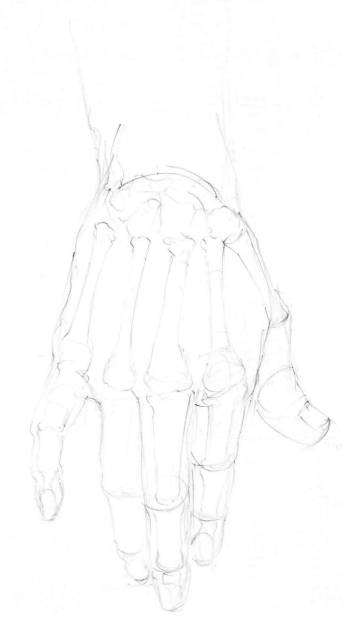

**Figure 90**   *Superimposition of bones into the hand.*

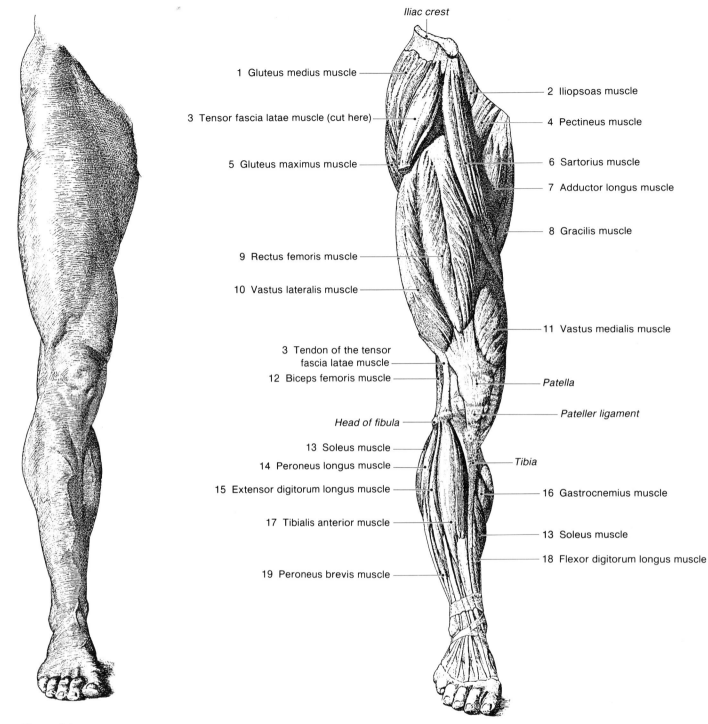

Iliac crest

1 Gluteus medius muscle

2 Iliopsoas muscle

3 Tensor fascia latae muscle (cut here)

4 Pectineus muscle

5 Gluteus maximus muscle

6 Sartorius muscle

7 Adductor longus muscle

8 Gracilis muscle

9 Rectus femoris muscle

10 Vastus lateralis muscle

11 Vastus medialis muscle

3 Tendon of the tensor
fascia latae muscle

12 Biceps femoris muscle

Patella

Pateller ligament

Head of fibula

13 Soleus muscle

14 Peroneus longus muscle

Tibia

15 Extensor digitorum longus muscle

16 Gastrocnemius muscle

17 Tibialis anterior muscle

13 Soleus muscle

18 Flexor digitorum longus muscle

19 Peroneus brevis muscle

**Figure 91** *Surface leg, anterior (front). As appeared in* An Atlas of Anatomy for Artists. *By Fritz Schider. Dover Publications, Inc. NY.*

**Figure 92** *Flayed leg, anterior (front). As appeared in* An Atlas of Anatomy for Artists. *By Fritz Schider. Dover Publications, Inc. NY.*

| Name | Action | Origin | Insertion |
|------|--------|--------|-----------|
| 1. GLUTEUS MEDIUS | Moves leg from standing to outstretched side position | Ilium | Femur |
| 2. ILIOPSOAS | Bends thigh at hip; rotates thigh outward; curls spine forward | Ilium & lower 4–5 vertebrae area | Femur |
| 3. TENSOR FASCIA LATAE | Rotates thigh inward; bends hip; moves thigh away from body | Anterior iliac spine | Iliotibial band |
| 4. PECTINEUS | Rotates thigh outward; bends thigh at hip; closes thigh from outstretched position | Pubis | Femur |
| 5. GLUTEUS MAXIMUS | Moves thigh backward; returns leg from side to standing position; rotates leg outward; presses buttocks together | Iliac crest, sacrum, & coccyx | Shaft of femur |
| 6. SARTORIUS | Bends thigh at hip; bends knee; moves thigh away from body; rotates thigh outward; rotates lower leg inward | Iliac spine | Tibia |
| 7. ADDUCTOR LONGUS | Bends thigh at hip; closes thigh from outstretched position; rotates thigh outward | Pubis | Femur |
| 8. GRACILIS | Bends thigh at hip; bends knee; closes thigh from outstretched position | Pubis | Tibia |
| 9. RECTUS FEMORIS | Straightens knee; helps to bend thigh at hip; moves thigh away from body at hip | Iliac spine | Patella & patellar ligament |
| 10. VASTUS LATERALIS | Same as RECTUS FEMORIS | Femur | Patella & patellar ligament |
| 11. VASTUS MEDIALIS | Same as RECTUS FEMORIS | Femur | Patella & patellar ligament |
| 12. BICEPS FEMORIS | Straightens thigh backward; rotates thigh outward; bends knee; rotates lower leg outward | Femur | Fibula |
| 13. SOLEUS | Points foot; turns soles of feet toward each other; "pigeon-toes" feet | Fibula & tibia | Achilles tendon to calcaneus |
| 14. PERONEUS LONGUS | Points foot; turns soles of feet away from each other | Fibula & tibia | Metatarsal and cuneiform of toe #1 |
| 15. EXTENSOR DIGITORUM LONGUS | Raises foot at ankle joint; raises toes #2–5 | Tibia & fibula | Tendons of toes #2–5 |
| 16. GASTROCNEMIUS | Points foot; bends knee; turns soles of feet toward each other; "pigeon-toes" feet | Femur | Achilles tendon to calcaneus |
| 17. TIBIALIS ANTERIOR | Raises foot; turns soles of feet toward each other | Tibia | Metatarsal and cuneiform (of toe #1) |
| 18. FLEXOR DIGITORUM LONGUS | Bends toes #2–5 | Tibia | Phalanges of toes #2–5 |
| 19. PERONEUS BREVIS | Points foot; turns soles of feet away from each other | Fibula | Metatarsal of toe #5 |

All actions and descriptive terms are defined with reference to the ANATOMICAL POSITION, that is, a figure standing erect, feet parallel and flat on the floor, eyes directed forward, arms at the sides, and palms facing forward.

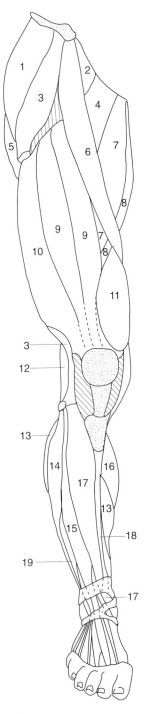

On pelvic girdle and/or vertebral column:
$O_{1,2,3,4,5,6,7,8,9,12}$

On tarsals, cuneiforms, metarsals, phalanges:
$I_{14,15,17,18,19}$

On iliotibial tract:
$I_{3,5}$

On calcaneus:
$I_{13,16}$

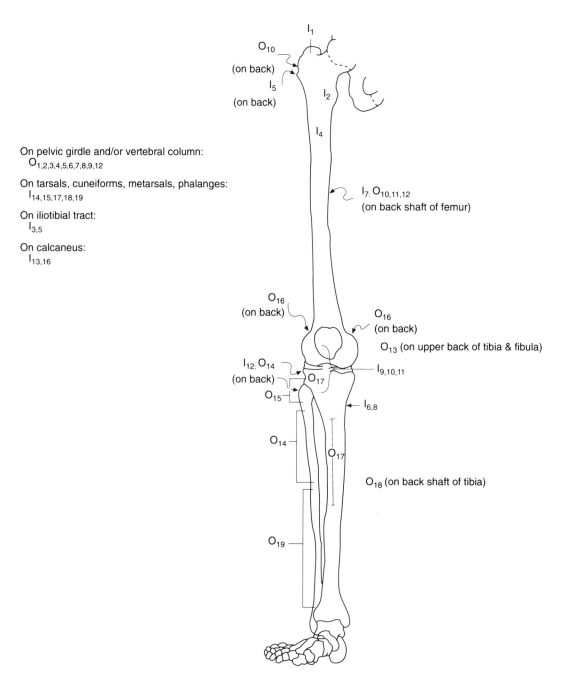

**Figure 93** *Coloring book of leg, anterior (front). As appeared in* An Atlas of Anatomy for Artists. By Fritz Schider. Dover Publications, Inc. NY.

**Figure 94** *Albinus (1697–1770). Leg bones, anterior (front). [Albinus on Anatomy. By Robert Beverly Hale and Terrance Coyle. Watson-Guptill Pub., 1979. p. 28.]*

# LEG—ANTERIOR

| Name | Action | Origin | Insertion |
|---|---|---|---|
| 1. GLUTEUS MEDIUS | Moves leg from standing to outstretched side position | Ilium | Femur |
| 2. ILIOPSOAS | Bends thigh at hip; rotates thigh outward; curls spine forward | Ilium & lower 4–5 vertebrae area | Femur |
| 3. TENSOR FASCIA LATAE | Rotates thigh inward; bends hip; moves thigh away from body | Anterior iliac spine | Iliotibial band |
| 4. PECTINEUS | Rotates thigh outward; bends thigh at hip; closes thigh from outstretched position | Pubis | Femur |
| 5. GLUTEUS MAXIMUS | Moves thigh backward; returns leg from side to standing position; rotates leg outward; presses buttocks together | Iliac crest, sacrum, & coccyx | Shaft of femur |
| 6. SARTORIUS | Bends thigh at hip; bends knee; moves thigh away from body; rotates thigh outward; rotates lower leg inward | Iliac spine | Tibia |
| 7. ADDUCTOR LONGUS | Bends thigh at hip; closes thigh from outstretched position; rotates thigh outward | Pubis | Femur |
| 8. GRACILIS | Bends thigh at hip; bends knee; closes thigh from outstretched position | Pubis | Tibia |
| 9. RECTUS FEMORIS | Straightens knee; helps to bend thigh at hip; moves thigh away from body at hip | Iliac spine | Patella & patellar ligament |
| 10. VASTUS LATERALIS | Same as RECTUS FEMORIS | Femur | Patella & patellar ligament |
| 11. VASTUS MEDIALIS | Same as RECTUS FEMORIS | Femur | Patella & patellar ligament |
| 12. BICEPS FEMORIS | Straightens thigh backward; rotates thigh outward; bends knee; rotates lower leg outward | Femur | Fibula |
| 13. SOLEUS | Points foot; turns soles of feet toward each other; "pigeon-toes" feet | Fibula & tibia | Achilles tendon to calcaneus |
| 14. PERONEUS LONGUS | Points foot; turns soles of feet away from each other | Fibula & tibia | Metatarsal and cuneiform of toe #1 |
| 15. EXTENSOR DIGITORUM LONGUS | Raises foot at ankle joint; raises toes #2–5 | Tibia & fibula | Tendons of toes #2–5 |
| 16. GASTROCNEMIUS | Points foot; bends knee; turns soles of feet toward each other; "pigeon-toes" feet | Femur | Achilles tendon to calcaneus |
| 17. TIBIALIS ANTERIOR | Raises foot; turns soles of feet toward each other | Tibia | Metatarsal and cuneiform (of toe #1) |
| 18. FLEXOR DIGITORUM LONGUS | Bends toes #2–5 | Tibia | Phalanges of toes #2–5 |
| 19. PERONEUS BREVIS | Points foot; turns soles of feet away from each other | Fibula | Metatarsal of toe #5 |

All actions and descriptive terms are defined with reference to the ANATOMICAL POSITION, that is, a figure standing erect, feet parallel and flat on the floor, eyes directed forward, arms at the sides, and palms facing forward.

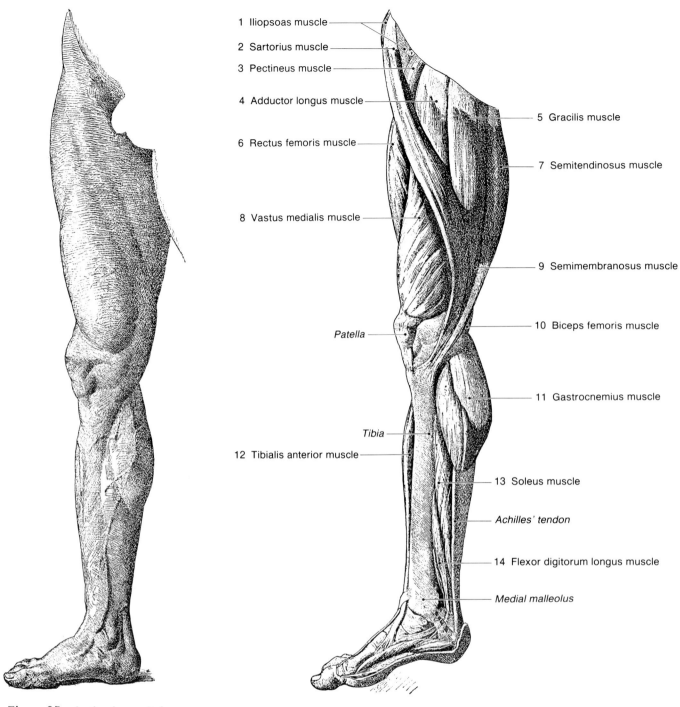

1 Iliopsoas muscle
2 Sartorius muscle
3 Pectineus muscle
4 Adductor longus muscle
5 Gracilis muscle
6 Rectus femoris muscle
7 Semitendinosus muscle
8 Vastus medialis muscle
9 Semimembranosus muscle
10 Biceps femoris muscle
*Patella*
11 Gastrocnemius muscle
*Tibia*
12 Tibialis anterior muscle
13 Soleus muscle
*Achilles' tendon*
14 Flexor digitorum longus muscle
*Medial malleolus*

***Figure 95*** *Surface leg, medial (inner).* As appeared in An Atlas of Anatomy for Artists. *By Fritz Schider. Dover Publications, Inc. NY.*

***Figure 96*** *Flayed leg, medial (inner).* As appeared in An Atlas of Anatomy for Artists. *By Fritz Schider. Dover Publications, Inc. NY.*

| Name | Action | Origin | Insertion |
|---|---|---|---|
| 1. ILIOPSOAS | Bends thigh at hip; rotates thigh outward; curls spine forward | Ilium & lower 4–5 vertebrae | Femur |
| 2. SARTORIUS | Bends thigh at hip; bends knee; moves thigh away from body; rotates thigh outward; rotates lower leg inward | Iliac spine | Tibia |
| 3. PECTINEUS | Rotates thigh outward; bends thigh at hip; closes thigh from outstretched position | Pubis | Femur |
| 4. ADDUCTOR LONGUS | Bends thigh at hip; closes thigh from outstretched position; rotates thigh outward | Pubis | Femur |
| 5. GRACILIS | Bends thigh at hip; bends knee; closes thigh from outstretched position | Pubis | Tibia |
| 6. RECTUS FEMORIS | Straightens knee; helps to bend thigh; moves thigh away from body | Iliac spine | Patella & patellar ligament |
| 7. SEMITENDINOSUS | Straightens thigh backward; closes thigh from outstretched position; rotates thigh inward; bends lower leg; rotates leg inward | Ischial tuberosity of pelvis | Tibia |
| 8. VASTUS MEDIALIS | Same as RECTUS FEMORIS | Femur | Patella & patellar ligament |
| 9. SEMIMEMBRANOSUS | Same as SEMITENDINOSUS | Ischial tuberosity of pelvis | Tibia |
| 10. BICEPS FEMORIS | Straightens thigh backward; rotates thigh outward; bends knee; rotates lower leg outward | Femur | Fibula |
| 11. GASTROCNEMIUS | Points foot; turns soles of feet toward each other; "pigeon-toes" feet | Femur | Achilles' tendon to calcaneus |
| 12. TIBIALIS ANTERIOR | Raises foot; turns soles of feet toward each other | Tibia | Metatarsal and cuneiform of toe #1 |
| 13. SOLEUS | Points foot; turns soles of feet toward each other; "pigeon-toes" feet | Fibula & tibia | Achilles' tendon to calcaneus |
| 14. FLEXOR DIGITORUM LONGUS | Bends toes #2–5 | Tibia | Phalanges of toes #2–5 |

All actions and descriptive terms are defined with reference to the ANATOMICAL POSITION, that is, a figure standing erect, feet parallel and flat on the floor, eyes directed forward, arms at the sides, and palms facing forward.

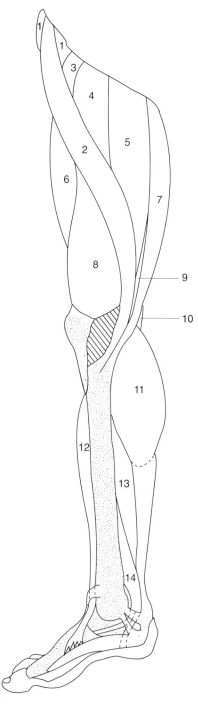

**Figure 97** *Coloring book of leg, medial (inner). As appeared in* An Atlas of Anatomy for Artists. *By Fritz Schider. Dover Publications, Inc. NY.*

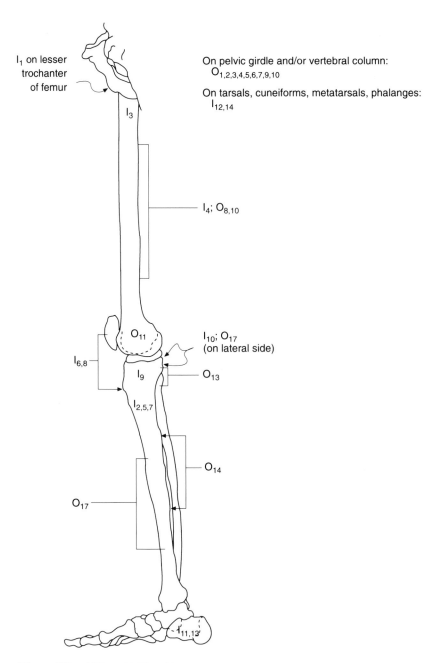

$I_1$ on lesser trochanter of femur

On pelvic girdle and/or vertebral column:
$O_{1,2,3,4,5,6,7,9,10}$

On tarsals, cuneiforms, metatarsals, phalanges:
$I_{12,14}$

$I_3$

$I_4$; $O_{8,10}$

$O_{11}$

$I_{10}$; $O_{17}$ (on lateral side)

$I_{6,8}$

$I_9$

$O_{13}$

$I_{2,5,7}$

$O_{14}$

$O_{17}$

$I_{11,13}$

**Figure 98** *Albinus (1697–1770). Leg bones, medial (inner). [Albinus on* Anatomy. *By Robert Beverly Hale and Terrance Coyle. Watson-Guptill Pub., NY. 1979. p. 32.]*

| Name | Action | Origin | Insertion |
|---|---|---|---|
| 1. ILIOPSOAS | Bends thigh at hip; rotates thigh outward; curls spine forward | Ilium & lower 4–5 vertebrae | Femur |
| 2. SARTORIUS | Bends thigh at hip; bends knee; moves thigh away from body; rotates thigh outward; rotates lower leg inward | Iliac spine | Tibia |
| 3. PECTINEUS | Rotates thigh outward; bends thigh at hip; closes thigh from outstretched position | Pubis | Femur |
| 4. ADDUCTOR LONGUS | Bends thigh at hip; closes thigh from outstretched position; rotates thigh outward | Pubis | Femur |
| 5. GRACILIS | Bends thigh at hip; bends knee; closes thigh from outstretched position | Pubis | Tibia |
| 6. RECTUS FEMORIS | Straightens knee; helps to bend thigh; moves thigh away from body | Iliac spine | Patella & patellar ligament |
| 7. SEMITENDINOSUS | Straightens thigh backward; closes thigh from outstretched position; rotates thigh inward; bends lower leg; rotates leg inward | Ischial tuberosity of pelvis | Tibia |
| 8. VASTUS MEDIALIS | Same as RECTUS FEMORIS | Femur | Patella & patellar ligament |
| 9. SEMIMEMBRANOSUS | Same as SEMITENDINOSUS | Ischial tuberosity of pelvis | Tibia |
| 10. BICEPS FEMORIS | Straightens thigh backward; rotates thigh outward; bends knee; rotates lower leg outward | Femur | Fibula |
| 11. GASTROCNEMIUS | Points foot; turns soles of feet toward each other; "pigeon-toes" feet | Femur | Achilles' tendon to calcaneus |
| 12. TIBIALIS ANTERIOR | Raises foot; turns soles of feet toward each other | Tibia | Metatarsal and cuneiform of toe #1 |
| 13. SOLEUS | Points foot; turns soles of feet toward each other; "pigeon-toes" feet | Fibula & tibia | Achilles' tendon to calcaneus |
| 14. FLEXOR DIGITORUM LONGUS | Bends toes #2–5 | Tibia | Phalanges of toes #2–5 |

All actions and descriptive terms are defined with reference to the
ANATOMICAL POSITION, that is, a figure standing erect, feet
parallel and flat on the floor, eyes directed forward, arms at the sides,
and palms facing forward.

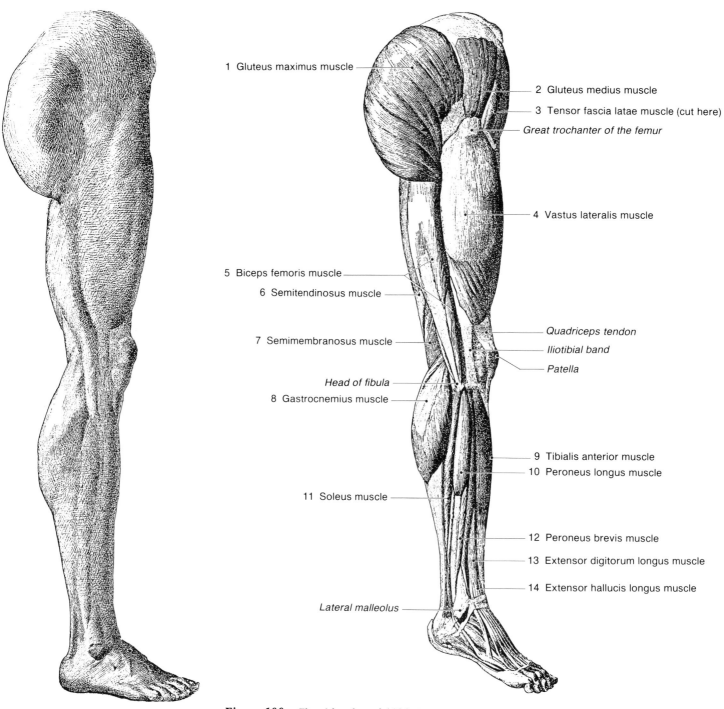

1 Gluteus maximus muscle

2 Gluteus medius muscle

3 Tensor fascia latae muscle (cut here)

*Great trochanter of the femur*

4 Vastus lateralis muscle

5 Biceps femoris muscle

6 Semitendinosus muscle

7 Semimembranosus muscle

*Quadriceps tendon*

*Iliotibial band*

*Patella*

*Head of fibula*

8 Gastrocnemius muscle

9 Tibialis anterior muscle

10 Peroneus longus muscle

11 Soleus muscle

12 Peroneus brevis muscle

13 Extensor digitorum longus muscle

14 Extensor hallucis longus muscle

*Lateral malleolus*

***Figure 99*** *Surface leg, lateral (side). As appeared in* An Atlas of Anatomy for Artists. *By Fritz Schider. Dover Publications, Inc. NY.*

***Figure 100*** *Flayed leg, lateral (side). As appeared in* An Atlas of Anatomy for Artists. *By Fritz Schider. Dover Publications, Inc. NY.*

| Name | Action | Origin | Insertion |
|---|---|---|---|
| 1. GLUTEUS MAXIMUS | Moves thigh backward; return leg from side to standing position; rotates leg outward; presses buttocks together | Iliac crest; sacrum & coccyx | Shaft of femur |
| 2. GLUTEUS MEDIUS | Moves leg from standing to outstretched side position | Ilium | Femur |
| 3. TENSOR FASCIA LATAE | Rotates thigh inward; bends hip; moves thigh away from body | Anterior iliac spine | Iliotibial band |
| 4. VASTUS LATERALIS | Same as RECTUS FEMORIS | Femur | Patella & patellar ligament |
| 5. BICEPS FEMORIS | Straightens thigh backward; rotates thigh outward; bends knee; rotates lower leg outward | Femur | Fibula |
| 6. SEMITENDINOSUS | Straightens thigh backward; closes thigh from outstretched position; rotates thigh inward; bends lower leg; rotates it inward | Ischial tuberosity of pelvis | Tibia |
| 7. SEMIMEMBRANOSUS | Same as SEMITENDINOSUS | Ischial tuberosity of pelvis | Tibia |
| 8. GASTROCNEMIUS | Points foot; bends knee; turns soles of feet toward each other; "pigeon-toes" feet | Femur | Achilles' tendon to calcaneus |
| 9. TIBIALIS ANTERIOR | Raises foot; turns soles of feet toward each other | Tibia | Metatarsal and cuneiform of toe #1 |
| 10. PERONEUS LONGUS | Points foot; turns soles of feet away from each other | Fibula & tibia | Metatarsal and cuneiform of toe #1 |
| 11. SOLEUS | Points foot; turns soles of feet toward each other; "pigeon-toes" feet | Fibula & tibia | Achilles' tendon to calcaneus |
| 12. PERONEUS BREVIS | Points foot; turn soles of feet away from each other | Fibula | Metatarsal of toe #5 |
| 13. EXTENSOR DIGITORUM LONGUS | Raises foot at ankle joint; raises toes #2–5 | Tibia & fibula | Tendons of toes #2–5 |
| 14. EXTENSOR HALLUCIS LONGUS | Helps to raise foot at ankle joint; raises big toe | Fibula | Last phalanx of big toe |

All actions and descriptive terms are defined with reference to the ANATOMICAL POSITION, that is, a figure standing erect, feet parallel and flat on the floor, eyes directed forward, arms at the sides, and palms facing forward.

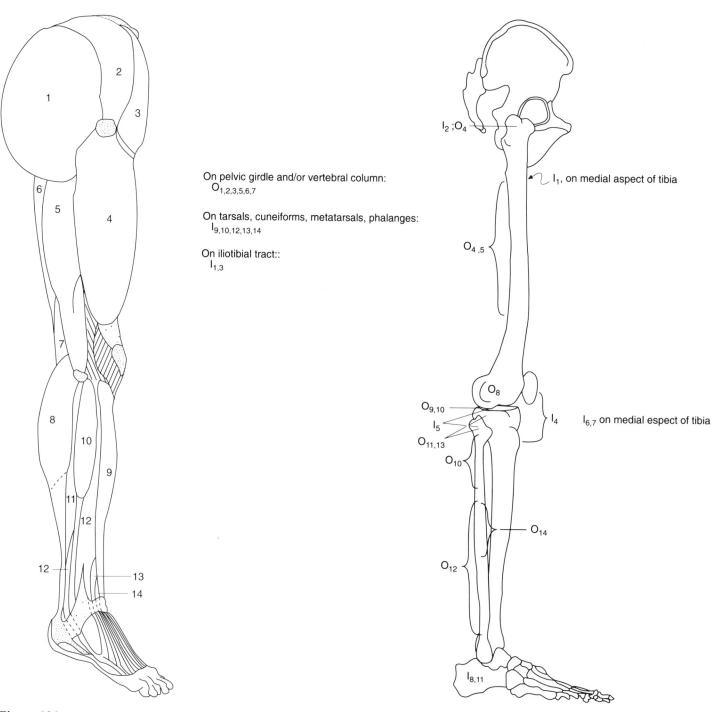

On pelvic girdle and/or vertebral column:
$O_{1,2,3,5,6,7}$

On tarsals, cuneiforms, metatarsals, phalanges:
$I_{9,10,12,13,14}$

On iliotibial tract::
$I_{1,3}$

$I_2 ; O_4$

$I_1$, on medial aspect of tibia

$O_{4,5}$

$O_8$

$O_{9,10}$

$I_5$

$O_{11,13}$

$O_{10}$

$I_4$

$I_{6,7}$ on medial espect of tibia

$O_{14}$

$O_{12}$

$I_{8,11}$

***Figure 101*** *Coloring book of leg, lateral (side).* As appeared in An Atlas of Anatomy for Artists. *By Fritz Schider. Dover Publications, Inc. NY.*

***Figure 102*** *Paul Richer. Leg bones, lateral (side).* [Artistic Anatomy. *By Paul Richer. R. B. Hale, translator. Watson-Guptill, NY. 1986.]*

| Name | Action | Origin | Insertion |
|------|--------|--------|-----------|
| 1. GLUTEUS MAXIMUS | Moves thigh backward; return leg from side to standing position; rotates leg outward; presses buttocks together | Iliac crest; sacrum & coccyx | Shaft of femur |
| 2. GLUTEUS MEDIUS | Moves leg from standing to outstretched side position | Ilium | Femur |
| 3. TENSOR FASCIA LATAE | Rotates thigh inward; bends hip; moves thigh away from body | Anterior iliac spine | Iliotibial band |
| 4. VASTUS LATERALIS | Same as RECTUS FEMORIS | Femur | Patella & patellar ligament |
| 5. BICEPS FEMORIS | Straightens thigh backward; rotates thigh outward; bends knee; rotates lower leg outward | Femur | Fibula |
| 6. SEMITENDINOSUS | Straightens thigh backward; closes thigh from outstretched position; rotates thigh inward; bends lower leg; rotates it inward | Ischial tuberosity of pelvis | Tibia |
| 7. SEMIMEMBRANOSUS | Same as SEMITENDINOSUS | Ischial tuberosity of pelvis | Tibia |
| 8. GASTROCNEMIUS | Points foot; bends knee; turns soles of feet toward each other; "pigeon-toes" feet | Femur | Achilles' tendon to calcaneus |
| 9. TIBIALIS ANTERIOR | Raises foot; turns soles of feet toward each other | Tibia | Metatarsal and cuneiform of toe #1 |
| 10. PERONEUS LONGUS | Points foot; turns soles of feet away from each other | Fibula & tibia | Metatarsal and cuneiform of toe #1 |
| 11. SOLEUS | Points foot; turns soles of feet toward each other; "pigeon-toes" feet | Fibula & tibia | Achilles' tendon to calcaneus |
| 12. PERONEUS BREVIS | Points foot; turn soles of feet away from each other | Fibula | Metatarsal of toe #5 |
| 13. EXTENSOR DIGITORUM LONGUS | Raises foot at ankle joint; raises toes #2–5 | Tibia & fibula | Tendons of toes #2–5 |
| 14. EXTENSOR HALLUCIS LONGUS | Helps to raise foot at ankle joint; raises big toe | Fibula | Last phalanx of big toe |

All actions and descriptive terms are defined with reference to the ANATOMICAL POSITION, that is, a figure standing erect, feet parallel and flat on the floor, eyes directed forward, arms at the sides, and palms facing forward.

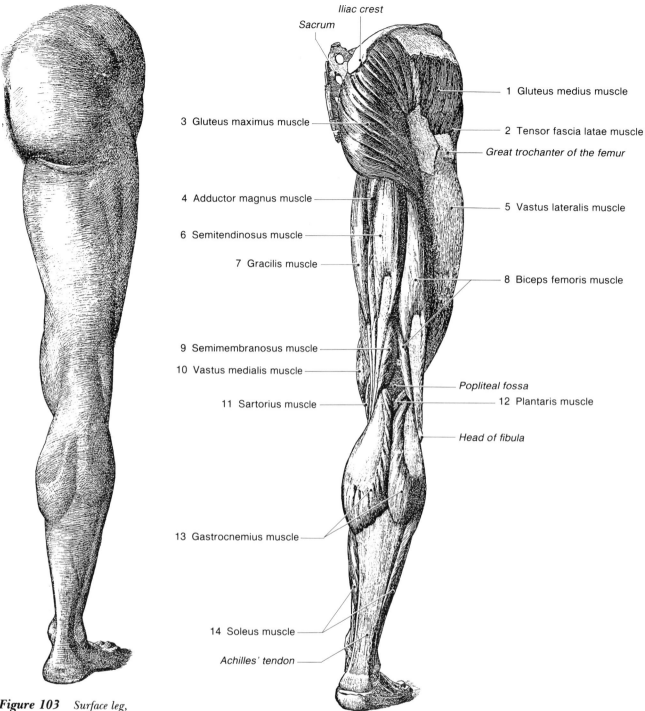

**Iliac crest**

**Sacrum**

1 Gluteus medius muscle

3 Gluteus maximus muscle

2 Tensor fascia latae muscle

*Great trochanter of the femur*

4 Adductor magnus muscle

5 Vastus lateralis muscle

6 Semitendinosus muscle

7 Gracilis muscle

8 Biceps femoris muscle

9 Semimembranosus muscle

10 Vastus medialis muscle

*Popliteal fossa*

11 Sartorius muscle

12 Plantaris muscle

*Head of fibula*

13 Gastrocnemius muscle

14 Soleus muscle

*Achilles' tendon*

***Figure 103*** *Surface leg, posterior (back). As appeared in* An Atlas of Anatomy for Artists. *By Fritz Schider. Dover Publications, Inc. NY.*

***Figure 104*** *Flayed leg, posterior (back). As appeared in* An Atlas of Anatomy for Artists. *By Fritz Schider. Dover Publications, Inc. NY.*

| Name | Action | Origin | Insertion |
|---|---|---|---|
| 1. GLUTEUS MEDIUS | Moves leg from standing to outstretched side position | Ilium | Femur |
| 2. TENSOR FASCIA LATAE | Rotates thigh inward; bends hip; moves thigh away from body | Anterior iliac spine | Iliotibial tract |
| 3. GLUTEUS MAXIMUS | Moves thigh backward; returns leg from side to standing position; rotates leg outward; presses buttocks together | Iliac crest, sacrum, & coccyx | Shaft of femur |
| 4. ADDUCTOR MAGNUS | Same as ADDUCTOR LONGUS | Ischium & pubis | Femur |
| 5. VASTUS LATERALIS | Same as RECTUS FEMORIS | Femur | Patella & patellar ligament |
| 6. SEMITENDINOSUS | Straightens thigh backward; closes thigh from outstretched position; rotates thigh inward; bends lower leg; rotates leg inward | Ischial tuberosity of pelvis | Tibia |
| 7. GRACILIS | Bends thigh at hip; bends knee; closes thigh from outstretched position | Pubis | Tibia |
| 8. BICEPS FEMORIS | Straightens thigh backward; rotates thigh outward; bends knee; rotates lower leg outward | Femur | Fibula |
| 9. SEMIMEMBRANOSUS | Same as SEMITENDINOSUS | Ischial tuberosity of pelvis | Tibia |
| 10. VASTUS MEDIALIS | Same as RECTUS FEMORIS | Femur | Patella & patellar ligament |
| 11. SARTORIUS | Bends thigh at hip; bends knee; moves thigh away from body; rotates thigh outward; rotates lower leg inward | Iliac spine | Tibia |
| 12. PLANTARIS | Points foot; bends knee joint | Femur | Calcaneus |
| 13. GASTROCNEMIUS | Points foot; bends knee; turns soles of feet toward each other; "pigeon-toes" feet | Femur | Achilles' tendon to calcaneus |
| 14. SOLEUS | Points foot; turns soles of feet toward each other; "pigeon-toes" feet | Fibula & tibia | Achilles' tendon to calcaneus |

All actions and descriptive terms are defined with reference to the ANATOMICAL POSITION, that is, a figure standing erect, feet parallel and flat on the floor, eyes directed forward, arms at the sides, and palms facing forward.

1
2
3
4
5
6
7
8
9
10
11
12
13
14

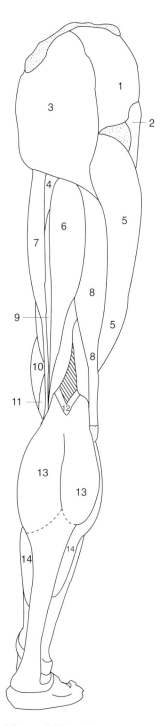

*Figure 105* *Coloring book of leg, posterior (back). As appeared in* An Atlas of Anatomy for Artists. *By Fritz Schider. Dover Publications, Inc. NY.*

On pelvic girdle and/or vertebral column:
$O_{1,2,3,4,6,7,8,9,11}$

On illiotibial tract:
$I_2$

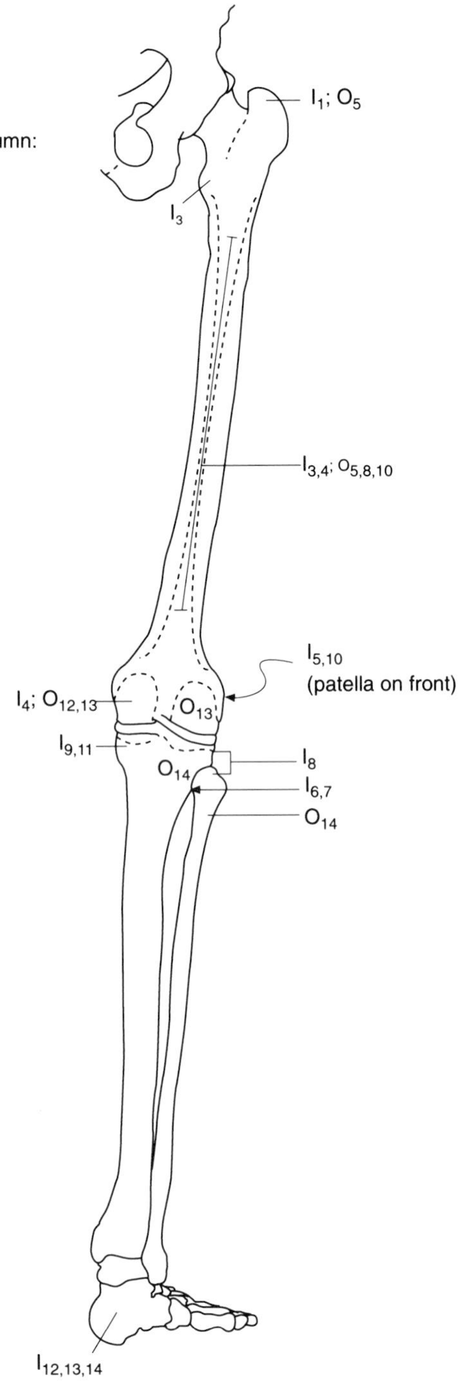

$I_1$; $O_5$
$I_3$
$I_{3,4}$; $O_{5,8,10}$
$I_{5,10}$
(patella on front)
$I_4$; $O_{12,13}$
$O_{13}$
$I_{9,11}$
$I_8$
$O_{14}$
$I_{6,7}$
$O_{14}$
$I_{12,13,14}$

*Figure 106* *Albinus (1697–1770). Leg bones, posterior (back). [*Albinus on Anatomy. *By Robert Beverly Hale and Terrance Coyle. Watson-Guptill Pub., NY. 1979. p. 30.]*

| Name | Action | Origin | Insertion |
|---|---|---|---|
| 1. GLUTEUS MEDIUS | Moves leg from standing to outstretched side position | Ilium | Femur |
| 2. TENSOR FASCIA LATAE | Rotates thigh inward; bends hip; moves thigh away from body | Anterior iliac spine | Iliotibial tract |
| 3. GLUTEUS MAXIMUS | Moves thigh backward; returns leg from side to standing position; rotates leg outward; presses buttocks together | Iliac crest, sacrum, & coccyx | Shaft of femur |
| 4. ADDUCTOR MAGNUS | Same as ADDUCTOR LONGUS | Ischium & pubis | Femur |
| 5. VASTUS LATERALIS | Same as RECTUS FEMORIS | Femur | Patella & patellar ligament |
| 6. SEMITENDINOSUS | Straightens thigh backward; closes thigh from outstretched position; rotates thigh inward; bends lower leg; rotates leg inward | Ischial tuberosity of pelvis | Tibia |
| 7. GRACILIS | Bends thigh at hip; bends knee; closes thigh from outstretched position | Pubis | Tibia |
| 8. BICEPS FEMORIS | Straightens thigh backward; rotates thigh outward; bends knee; rotates lower leg outward | Femur | Fibula |
| 9. SEMIMEMBRANOSUS | Same as SEMITENDINOSUS | Ischial tuberosity of pelvis | Tibia |
| 10. VASTUS MEDIALIS | Same as RECTUS FEMORIS | Femur | Patella & patellar ligament |
| 11. SARTORIUS | Bends thigh at hip; bends knee; moves thigh away from body; rotates thigh outward; rotates lower leg inward | Iliac spine | Tibia |
| 12. PLANTARIS | Points foot; bends knee joint | Femur | Calcaneus |
| 13. GASTROCNEMIUS | Points foot; bends knee; turns soles of feet toward each other; "pigeon-toes" feet | Femur | Achilles' tendon to calcaneus |
| 14. SOLEUS | Points foot; turns soles of feet toward each other; "pigeon-toes" feet | Fibula & tibia | Achilles' tendon to calcaneus |

All actions and descriptive terms are defined with reference to the ANATOMICAL POSITION, that is, a figure standing erect, feet parallel and flat on the floor, eyes directed forward, arms at the sides, and palms facing forward.

## Legs

Position yourself to the skeleton, front, side or back. The legs of the skeleton, for this exercise, should remain vertical to the floor. Again on separate sheets of paper, draw the blind contour, then the gestural sighting with geometric volumes. The third drawing is the fleshy outline of the model's leg. Remember to take time to draw each of these drawings separately and carefully. They are teaching tools. You are teaching yourself to observe. You are learning a great deal. Repetition through these exercises will enhance your work each time. Sudden breakthroughs sometimes come as you draw. You may find the task of drawing no longer a task, but a compelling adventure. Keep on going. Try to name the bones as you go, or do that before you begin to draw. Note how, in both the arm and leg, the rotation of the bones on the lower part of the limb allows for extraordinary changes in musculature. We will learn more of that very soon. ▲

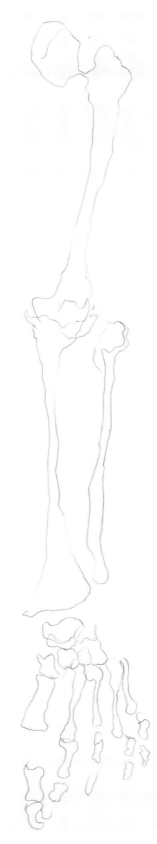

***Figure 107***   *Blind contour of a leg.*

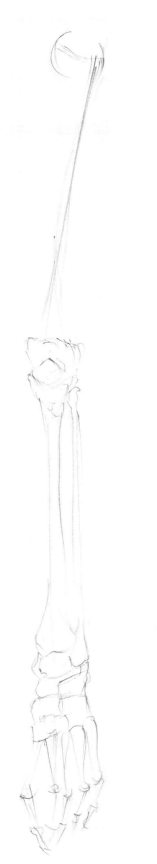

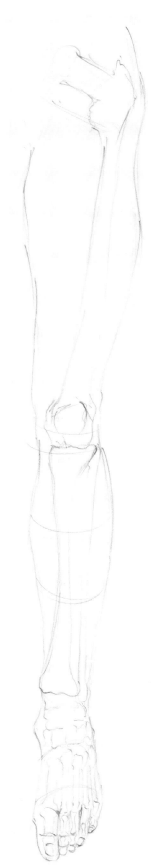

**Figure 108**    *Sighting, gesture, and volumes of a leg.*

**Figure 109**    *Bones superimposed into the leg.*

107

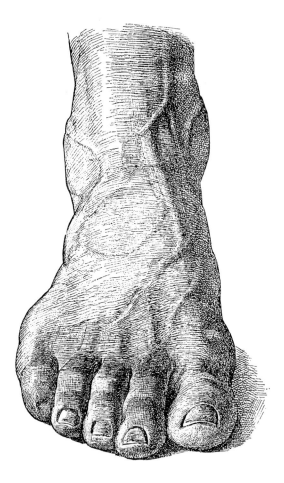

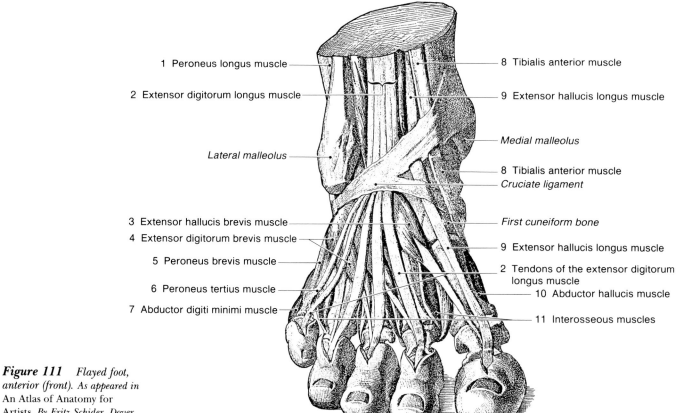

1 Peroneus longus muscle

2 Extensor digitorum longus muscle

Lateral malleolus

3 Extensor hallucis brevis muscle
4 Extensor digitorum brevis muscle
5 Peroneus brevis muscle
6 Peroneus tertius muscle
7 Abductor digiti minimi muscle

8 Tibialis anterior muscle

9 Extensor hallucis longus muscle

Medial malleolus

8 Tibialis anterior muscle
Cruciate ligament

First cuneiform bone

9 Extensor hallucis longus muscle

2 Tendons of the extensor digitorum longus muscle

10 Abductor hallucis muscle

11 Interosseous muscles

FOOT—ANTERIOR

| Name | Action | Origin | Insertion |
|---|---|---|---|
| 1. PERONEUS LONGUS | Points foot; turns soles of feet away from each other | Fibula & tibia | Metatarsal and cuneiform of toe #1 |
| 2. EXTENSOR DIGITORUM LONGUS | Raises foot at ankle joint; raises toes #2–5 | Tibia & fibula | Tendons of toes #2–5 |
| 3. EXTENSOR HALLUCIS BREVIS | Helps to raise foot at ankle joint; raises big toe | Fibula | First phalanx of big toe |
| 4. EXTENSOR DIGITORUM BREVIS | Raises foot at ankle joint; raises toes #2–5 | Tibia & fibula | Tendons of toes #2–5 |
| 5. PERONEUS BREVIS | Points foot; turns soles of feet away from each other | Fibula | Metatarsal of toe #5 |
| 6. PERONEUS TERTIUS | Turns foot out at heel; raises foot; turns soles of feet away from each other; raises toes II, III, IV, V | Extensor digitorum longus | Metatarsal #5 |
| 7. ABDUCTOR DIGITI MINIMI | Curls little toe downward and outward | Calcaneus | Metatarsal #5 and first phalanx of #5 |
| 8. TIBIALIS ANTERIOR | Raises foot; turns soles of feet toward each other | Tibia | Metatarsal and cuneiform of toe #1 |
| 9. EXTENSOR HALLUCIS LONGUS | Helps to raise foot at ankle joint; raises big toe | Fibula | Last phalanx of big toe |
| 10. ABDUCTOR HALLUCIS | Bends big toe downward; draws big toe away from others | Tibia; calcaneus | First phalanx of big toe |
| 11. INTEROSSEI MUSCLES | Bends toes; spreads toes apart | Adjacent sides of metatarsal bones | Phalanges II, III, IV |

All actions and descriptive terms are defined with reference to the ANATOMICAL POSITION, that is, a figure standing erect, feet parallel and flat on the floor, eyes directed forward, arms at the sides, and palms facing forward.

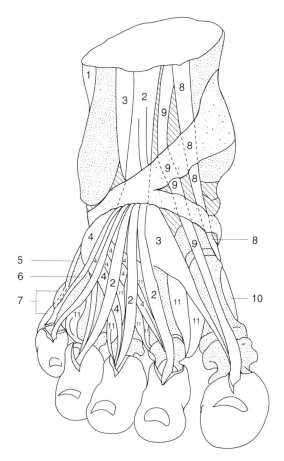

On tibia and/or fibula:
$O_{1,2,5,6,8,9,10}$

On tendons:
$I_{2,4,}$

On calcaneus:
$O_{3,4,7,10}$

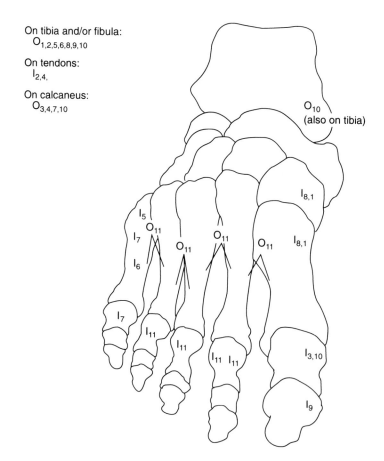

**Figure 112**   *Coloring book of foot, anterior (front). As appeared in* An Atlas of Anatomy for Artists. *By Fritz Schider. Dover Publications, Inc. NY.*

**Figure 113**   *Albinus (1697–1770). Foot bones, anterior (front).* [Albinus on Anatomy. *By Robert Beverly Hale and Terrance Coyle. Watson-Guptill Pub., NY. 1979. p. 142.]*

| Name | Action | Origin | Insertion |
|------|--------|--------|-----------|
| 1. PERONEUS LONGUS | Points foot; turns soles of feet away from each other | Fibula & tibia | Metatarsal and cuneiform of toe #1 |
| 2. EXTENSOR DIGITORUM LONGUS | Raises foot at ankle joint; raises toes #2–5 | Tibia & fibula | Tendons of toes #2–5 |
| 3. EXTENSOR HALLUCIS BREVIS | Helps to raise foot at ankle joint; raises big toe | Fibula | First phalanx of big toe |
| 4. EXTENSOR DIGITORUM BREVIS | Raises foot at ankle joint; raises toes #2–5 | Tibia & fibula | Tendons of toes #2–5 |
| 5. PERONEUS BREVIS | Points foot; turns soles of feet away from each other | Fibula | Metatarsal of toe #5 |
| 6. PERONEUS TERTIUS | Turns foot out at heel; raises foot; turns soles of feet away from each other; raises toes II, III, IV, V | Extensor digitorum longus | Metatarsal #5 |
| 7. ABDUCTOR DIGITI MINIMI | Curls little toe downward and outward | Calcaneus | Metatarsal #5 and first phalanx of #5 |
| 8. TIBIALIS ANTERIOR | Raises foot; turns soles of feet toward each other | Tibia | Metatarsal and cuneiform of toe #1 |
| 9. EXTENSOR HALLUCIS LONGUS | Helps to raise foot at ankle joint; raises big toe | Fibula | Last phalanx of big toe |
| 10. ABDUCTOR HALLUCIS | Bends big toe downward; draws big toe away from others | Tibia; calcaneus | First phalanx of big toe |
| 11. INTEROSSEI MUSCLES | Bends toes; spreads toes apart | Adjacent sides of metatarsal bones | Phalanges II, III, IV |

All actions and descriptive terms are defined with reference to the ANATOMICAL POSITION, that is, a figure standing erect, feet parallel and flat on the floor, eyes directed forward, arms at the sides, and palms facing forward.

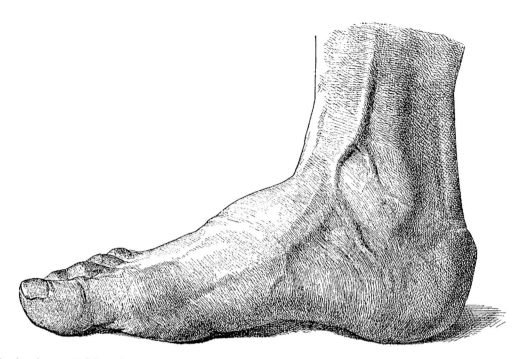

**Figure 114**    *Surface foot, medial (inner). As appeared in* An Atlas of Anatomy for Artists. *By Fritz Schider. Dover Publications, Inc. NY.*

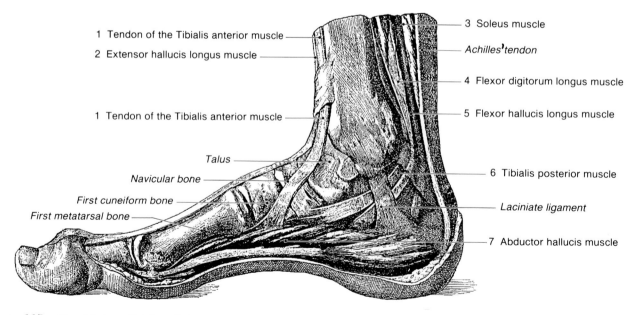

1 Tendon of the Tibialis anterior muscle

2 Extensor hallucis longus muscle

3 Soleus muscle

Achilles' tendon

4 Flexor digitorum longus muscle

5 Flexor hallucis longus muscle

1 Tendon of the Tibialis anterior muscle

Talus

Navicular bone

First cuneiform bone

First metatarsal bone

6 Tibialis posterior muscle

Laciniate ligament

7 Abductor hallucis muscle

**Figure 115**    *Flayed foot, medial (inner). As appeared in* An Atlas of Anatomy for Artists. *By Fritz Schider. Dover Publications, Inc. NY.*

| Name | Action | Origin | Insertion |
|---|---|---|---|
| 1. TIBIALIS ANTERIOR | Raises foot; turns soles of feet toward each other | Tibia | Metatarsal & cuneiform of toe #1 |
| 2. EXTENSOR HALLUCIS LONGUS | Helps to raise foot at ankle joint; raises big toe | Fibula | Last phalanx of big toe |
| 3. SOLEUS | Points foot; turns soles of feet toward each other; "pigeon-toes" feet | Fibula & tibia | Achilles' tendon to calcaneus |
| 4. FLEXOR DIGITORUM LONGUS | Bends toes #2–5 | Tibia | Phalanges of toes #2–5 |
| 5. FLEXOR HALLUCIS LONGUS | Points feet; turns soles of feet toward each other; bends big toe; "pigeon-toes" feet | Fibula | Big toe |
| 6. TIBIALIS POSTERIOR | Points feet; turns soles of feet toward each other; supports arch in foot | Fibula; tibia | Navicular bone |
| 7. ABDUCTOR HALLUCIS | Bends big toe downward; draws big toe away from others | Tibia; calcaneus | First phalanx of big toe |

All actions and descriptive terms are defined with reference to the ANATOMICAL POSITION, that is, a figure standing erect, feet parallel and flat on the floor, eyes directed forward, arms at the sides, and palms facing forward.

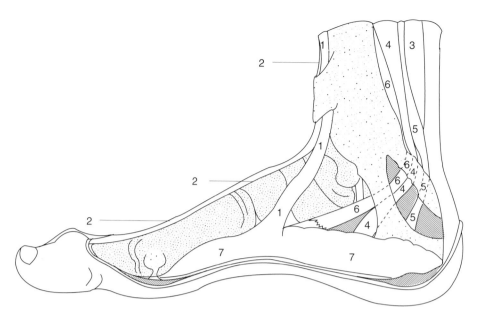

**Figure 116**  *Coloring book of foot, medial (inner). As appeared in* An Atlas of Anatomy for Artists. *By Fritz Schider. Dover Publications, Inc. NY.*

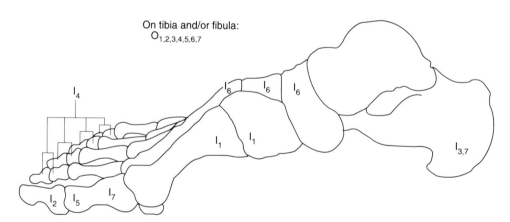

On tibia and/or fibula:
$O_{1,2,3,4,5,6,7}$

**Figure 117**  *Albinus (1697–1770). Foot bones, medial (inner).* [Albinus on Anatomy. *By Robert Beverly Hale and Terrance Coyle. Watson-Guptill Pub., NY. 1979. p. 144.]*

| Name | Action | Origin | Insertion |
|---|---|---|---|
| 1. TIBIALIS ANTERIOR | Raises foot; turns soles of feet toward each other | Tibia | Metatarsal & cuneiform of toe #1 |
| 2. EXTENSOR HALLUCIS LONGUS | Helps to raise foot at ankle joint; raises big toe | Fibula | Last phalanx of big toe |
| 3. SOLEUS | Points foot; turns soles of feet toward each other; "pigeon-toes" feet | Fibula & tibia | Achilles' tendon to calcaneus |
| 4. FLEXOR DIGITORUM LONGUS | Bends toes #2–5 | Tibia | Phalanges of toes #2–5 |
| 5. FLEXOR HALLUCIS LONGUS | Points feet; turns soles of feet toward each other; bends big toe; "pigeon-toes" feet | Fibula | Big toe |
| 6. TIBIALIS POSTERIOR | Points feet; turns soles of feet toward each other; supports arch in foot | Fibula; tibia | Navicular bone |
| 7. ABDUCTOR HALLUCIS | Bends big toe downward; draws big toe away from others | Tibia; calcaneus | First phalanx of big toe |

All actions and descriptive terms are defined with reference to the ANATOMICAL POSITION, that is, a figure standing erect, feet parallel and flat on the floor, eyes directed forward, arms at the sides, and palms facing forward.

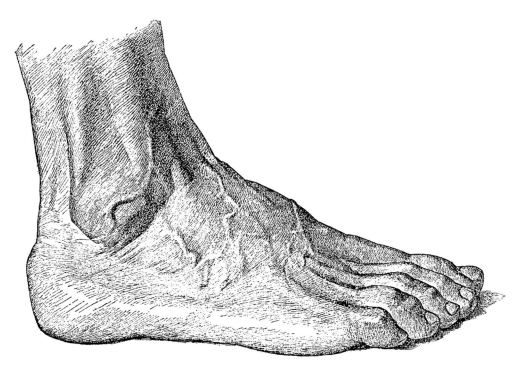

***Figure 118***    *Surface foot, lateral (side). As appeared in* An Atlas of Anatomy for Artists. *By Fritz Schider. Dover Publications, Inc. NY.*

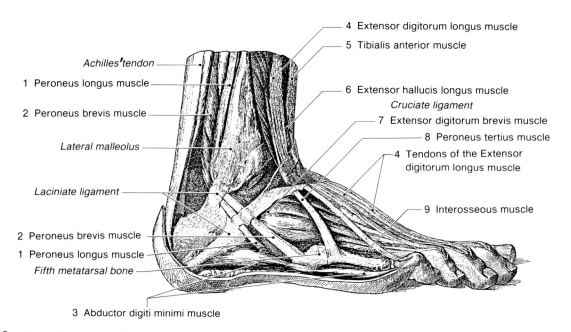

Achilles'tendon

1 Peroneus longus muscle

2 Peroneus brevis muscle

Lateral malleolus

Laciniate ligament

2 Peroneus brevis muscle

1 Peroneus longus muscle

Fifth metatarsal bone

3 Abductor digiti minimi muscle

4 Extensor digitorum longus muscle

5 Tibialis anterior muscle

6 Extensor hallucis longus muscle

Cruciate ligament

7 Extensor digitorum brevis muscle

8 Peroneus tertius muscle

4 Tendons of the Extensor digitorum longus muscle

9 Interosseous muscle

***Figure 119***    *Flayed foot, lateral (side). As appeared in* An Atlas of Anatomy for Artists. *By Fritz Schider. Dover Publications, Inc. NY.*

| Name | Action | Origin | Insertion |
|------|--------|--------|-----------|
| 1. PERONEUS LONGUS | Points foot; turns soles of feet toward each other | Fibula & tibia | Metatarsal and cuneiform of toe #1 |
| 2. PERONEUS BREVIS | Points foot; turns soles of feet away from each other | Fibula | Metatarsal of toe #5 |
| 3. ABDUCTOR DIGITI MINIMI | Curls little toe downward and outward | Calcaneus | Metatarsal V and first phalanx of V |
| 4. EXTENSOR DIGITORUM LONGUS | Raises foot at ankle joint; raises toes #2–5 | Tibia & fibula | Tendons of toes #2–5 |
| 5. TIBIALIS ANTERIOR | Raises foot; turns soles of feet toward each other | Tibia | Metatarsal and cuneiform of toe #1 |
| 6. EXTENSOR HALLUCIS LONGUS | Helps to raise foot at ankle joint; raises big toe | Fibula | Last phalanx of big toe |
| 7. EXTENSOR DIGITORUM BREVIS | Spreads apart toes #2–4; raises them up | Calcaneus | Tendons of toes #2–4 |
| 8. PERONEUS TERTIUS | Turns foot out at heel; raises foot; turns soles of feet away from each other; raises toes #2–5 | Extensor digitorum longus | Metatarsal #5 |
| 9. INTEROSSEI MUSCLES | Bends toes; spreads toes apart | Adjacent sides of metatarsal bones | Phalanges II, III, IV, V |

All actions and descriptive terms are defined with reference to the ANATOMICAL POSITION, that is, a figure standing erect, feet parallel and flat on the floor, eyes directed forward, arms at the sides, and palms facing forward.

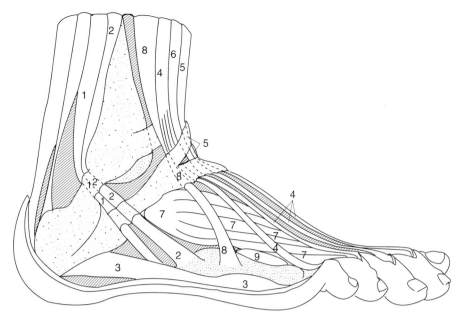

**Figure 120**  *Coloring book of foot, lateral (side). As appeared in* An Atlas of Anatomy for Artists. *By Fritz Schider. Dover Publications, Inc. NY.*

On tibia and/or fibula:
$O_{1,2,4,5,6,8}$

On tendon:
$I_4$

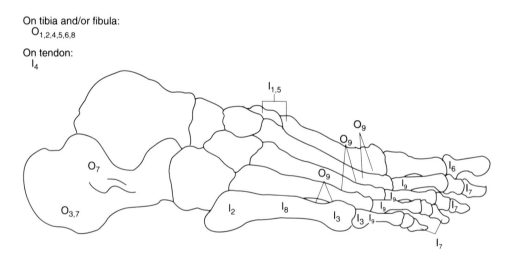

**Figure 121**  *Albinus (1697–1770). Foot bones, lateral (side). [*Albinus on Anatomy. *By Robert Beverly Hale and Terrance Coyle. Watson-Guptill Pub., NY. 1979. p. 144.]*

| Name | Action | Origin | Insertion |
|------|--------|--------|-----------|
| 1. PERONEUS LONGUS | Points foot; turns soles of feet toward each other | Fibula & tibia | Metatarsal and cuneiform of toe #1 |
| 2. PERONEUS BREVIS | Points foot; turns soles of feet away from each other | Fibula | Metatarsal of toe #5 |
| 3. ABDUCTOR DIGITI MINIMI | Curls little toe downward and outward | Calcaneus | Metatarsal V and first phalanx of V |
| 4. EXTENSOR DIGITORUM LONGUS | Raises foot at ankle joint; raises toes #2–5 | Tibia & fibula | Tendons of toes #2–5 |
| 5. TIBIALIS ANTERIOR | Raises foot; turns soles of feet toward each other | Tibia | Metatarsal and cuneiform of toe #1 |
| 6. EXTENSOR HALLUCIS LONGUS | Helps to raise foot at ankle joint; raises big toe | Fibula | Last phalanx of big toe |
| 7. EXTENSOR DIGITORUM BREVIS | Spreads apart toes #2–4; raises them up | Calcaneus | Tendons of toes #2–4 |
| 8. PERONEUS TERTIUS | Turns foot out at heel; raises foot; turns soles of feet away from each other; raises toes #2–5 | Extensor digitorum longus | Metatarsal #5 |
| 9. INTEROSSEI MUSCLES | Bends toes; spreads toes apart | Adjacent sides of metatarsal bones | Phalanges II, III, IV, V |

All actions and descriptive terms are defined with reference to the ANATOMICAL POSITION, that is, a figure standing erect, feet parallel and flat on the floor, eyes directed forward, arms at the sides, and palms facing forward.

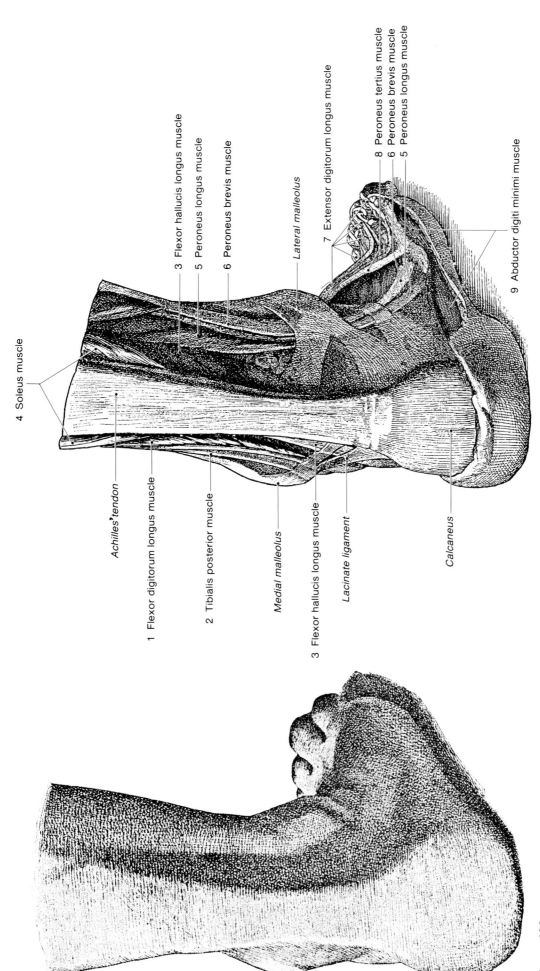

4 Soleus muscle

Achilles' tendon

1 Flexor digitorum longus muscle

2 Tibialis posterior muscle

Medial malleolus

3 Flexor hallucis longus muscle

Lacinate ligament

Calcaneus

3 Flexor hallucis longus muscle

5 Peroneus longus muscle

6 Peroneus brevis muscle

Lateral malleolus

7 Extensor digitorum longus muscle

8 Peroneus tertius muscle

6 Peroneus brevis muscle

5 Peroneus longus muscle

9 Abductor digiti minimi muscle

**Figure 123** *Flayed foot, posterior (back). As appeared in* An Atlas of Anatomy for Artists. *By Fritz Schider. Dover Publications, Inc. NY.*

**Figure 122** *Surface foot, posterior (back). As appeared in An Atlas of Anatomy for Artists. By Fritz Schider. Dover Publications, Inc. NY.*

| Name | Action | Origin | Insertion |
|------|--------|--------|-----------|
| 1. FLEXOR DIGITORUM LONGUS | Bends toes #2–5 | Tibia | Phalanges of toes #2–5 |
| 2. TIBIALIS POSTERIOR | Points feet; turns soles of feet toward each other; supports arch in foot | Fibula & tibia | Navicular bone |
| 3. FLEXOR HALLUCIS LONGUS | | Fibula | Big toe |
| 4. SOLEUS | Points foot; turns soles of feet toward each other; "pigeon-toes" feet | Fibula & tibia | Achilles' tendon to calcaneus |
| 5. PERONEUS LONGUS | Points foot; turns soles of feet toward each other | Fibula & tibia | Metatarsal & cuneiform of toe #1 |
| 6. PERONEUS BREVIS | Points foot; turns soles of feet away from each other | Fibula | Metatarsal of toe #5 |
| 7. EXTENSOR DIGITORUM LONGUS | Raises foot at ankle joint; raises toes #2–5 | Tibia & fibula | Tendons of toes #2–5 |
| 8. PERONEUS TERTIUS | Turns foot out at heel; raises foot; turns soles of feet away from each other; raises toes II, III, IV, V | Extensor digitorum longus | Metatarsal V |
| 9. ABDUCTOR DIGITI MINIMI | Curls little toe downward and outward | Calcaneus | Metatarsal V and first phalanx of V |
| 10. EXTENSOR DIGITORUM BREVIS | Spreads apart toes #2–4; raises them up | Calcaneus | Tendons of toes #2–4 |

All actions and descriptive terms are defined with reference to the ANATOMICAL POSITION, that is, a figure standing erect, feet parallel and flat on the floor, eyes directed forward, arms at the sides, and palms facing forward.

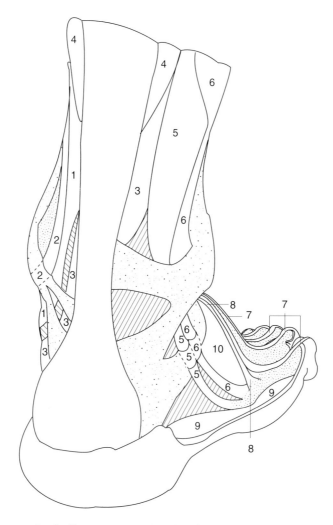

**Figure 124**  *Coloring book of foot, posterior (back).* As appeared in An Atlas of Anatomy for Artists. *By Fritz Schider. Dover Publications, Inc. NY.*

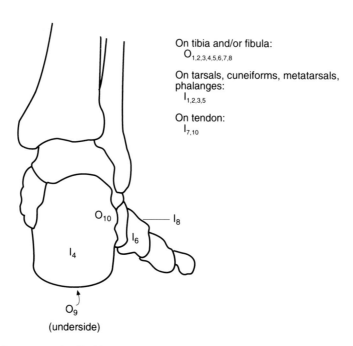

On tibia and/or fibula:
$O_{1,2,3,4,5,6,7,8}$

On tarsals, cuneiforms, metatarsals, phalanges:
$I_{1,2,3,5}$

On tendon:
$I_{7,10}$

**Figure 125**  *Paul Richer. Foot bones, posterior (back).* [Artistic Anatomy. By Dr. Paul Richer. Tanslated and edited by Robert Beverly Hale. Watson-Guptill Publ., NY. 1986.]

| Name | Action | Origin | Insertion |
|------|--------|--------|-----------|
| 1. FLEXOR DIGITORUM LONGUS | Bends toes #2–5 | Tibia | Phalanges of toes #2–5 |
| 2. TIBIALIS POSTERIOR | Points feet; turns soles of feet toward each other; supports arch in foot | Fibula & tibia | Navicular bone |
| 3. FLEXOR HALLUCIS LONGUS | | Fibula | Big toe |
| 4. SOLEUS | Points foot; turns soles of feet toward each other; "pigeon-toes" feet | Fibula & tibia | Achilles' tendon to calcaneus |
| 5. PERONEUS LONGUS | Points foot; turns soles of feet toward each other | Fibula & tibia | Metatarsal & cuneiform of toe #1 |
| 6. PERONEUS BREVIS | Points foot; turns soles of feet away from each other | Fibula | Metatarsal of toe #5 |
| 7. EXTENSOR DIGITORUM LONGUS | Raises foot at ankle joint; raises toes #2–5 | Tibia & fibula | Tendons of toes #2–5 |
| 8. PERONEUS TERTIUS | Turns foot out at heel; raises foot; turns soles of feet away from each other; raises toes II, III, IV, V | Extensor digitorum longus | Metatarsal V |
| 9. ABDUCTOR DIGITI MINIMI | Curls little toe downward and outward | Calcaneus | Metatarsal V and first phalanx of V |
| 10. EXTENSOR DIGITORUM BREVIS | Spreads apart toes #2–4; raises them up | Calcaneus | Tendons of toes #2–4 |

All actions and descriptive terms are defined with reference to the ANATOMICAL POSITION, that is, a figure standing erect, feet parallel and flat on the floor, eyes directed forward, arms at the sides, and palms facing forward.

## Feet

Feet often seem difficult to draw because, like hands, not much mass is apparent compared to the shoulders, arms, or buttocks. Feet and hands drawn too quickly appear like flippers, slippers, or mittens. If that shows in your drawing, you are working too quickly. Hands and feet deserve respect. They have their own uniquenesses. So, take the time to draw two feet, preferably in different postures, in blind contour, first. Secondly, in gesture with sighting and volumes. Thirdly, draw the outline of the model's foot. Carefully sight from the skeleton, the heel bone to ankle. The heel protrudes surprisingly. Impose the bones into the fleshy outline of the model's foot. Use the book for reference or your own foot if you need it to see just exactly where muscle and sinew rise and fall along the apparent edges of the bone structure. ▲

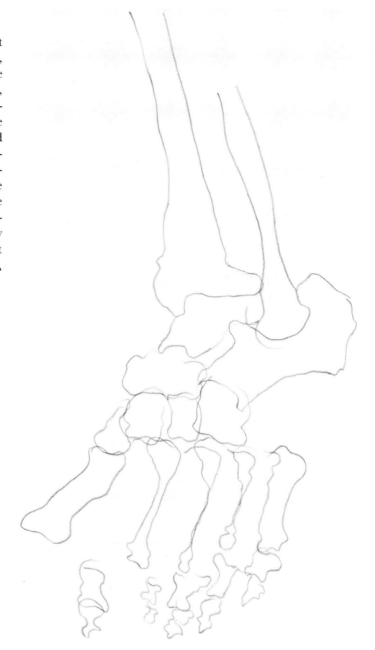

***Figure 126*** *Blind contour of a foot.*

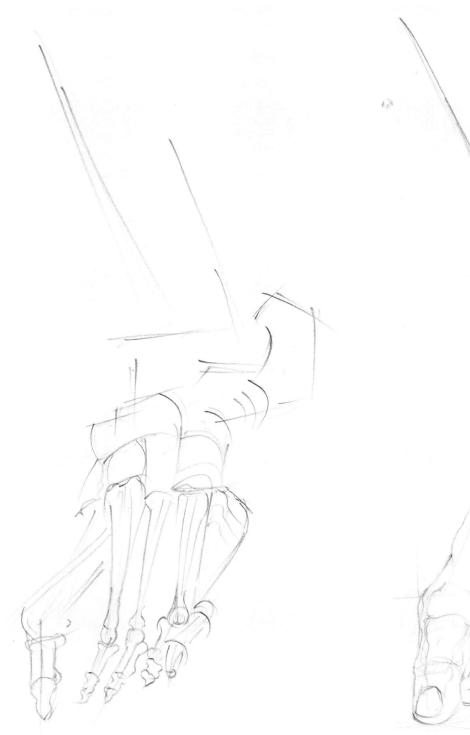

**Figure 127**  *Sighting, gesture, and volumes of a foot.*

**Figure 128**  *Bones superimposed into the flesh of a foot.*

## Muscles

Muscles are connected with bones, cartilages, ligaments, and skin directly, or, with tendons, indirectly. Muscles vary considerably in their own form. In the limbs they surround the bones. In the trunk they are broad and flat, encasing cavities that hold organs.

Fibers in individual muscles vary too. Some fibers are parallel from origin to insertion. Some curve and taper at the ends. Some converge at the insertion. Some fibers are at diagonal and insert that way along a tendon. And some are arranged in bundles.

The arrangement of muscle fibers is important to their function and movement. Muscles with fibers that are long have diminished strength, and muscles with shorter and more numerous fibers have more power.

Muscles differ much in size. The fibers of the Sartorius are nearly two feet in length. Inside the ear one can find a muscle about two pencil lines wide.

Naming muscles and remembering the names helps if you understand their derivation. "The names applied to various muscles have been derived (1) from their *SITUATION,* as the Tibialis, Radialias, Ulnaris, Peroneus; (2) from their *DIRECTION,* as the Rectus abdominis, Obliqui capitis, Transversalis; (3) from their *USES,* as Flexors, Extensors, Abductors, etc.; (4) from their *SHAPE,* as the Deltoid, Trapesius, Rhomboideus; (5) from the *NUMBER of their DIVISIONS,* as the Biceps, the Triceps; and (6) from their *POINT OF ATTACHMENT,* as the Sterno-cleido-mastoid, Sterno-hyoid, Sterno-thyroid."[1] Understanding the action of a muscle of necessity means that you understand the points of attachment.

Within this section of the anatomical study are specialized exercises. In some cases you might need to refer back into the coloring book section for origin and insertion of the superficial muscles. Some of the following illustrations suggest tracing sheet overlays to color code muscles, most especially the facial muscles. Other illustrations are simply here for you to study. However, in almost all cases I would suggest you position yourself to the model to duplicate what you see in the book illustration, then: (1) draw the section you are studying from the model, (2) superimpose the muscles. Try to think sighting and volume. Referring to the book or a plaster model of a splayed figure you might have in your room could help you. ▲

Icularis oculi
Omaticus minor
Omaticus major
Seter
Icularis oris
Angularis
Italis

***Figure 129*** *Jean Galbert Savage. Muscles of the head. Courtesy of the Francis A. Countway Library of Medicine.*

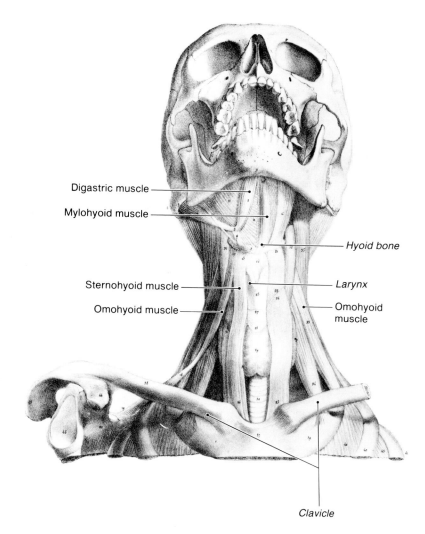

Digastric muscle

Mylohyoid muscle

Hyoid bone

Sternohyoid muscle

Larynx

Omohyoid muscle

Omohyoid muscle

Clavicle

**Figure 130**  *Raised skull with muscles of the neck. Lithograph. Courtesy of the Francis A. Countway Library of Medicine.*

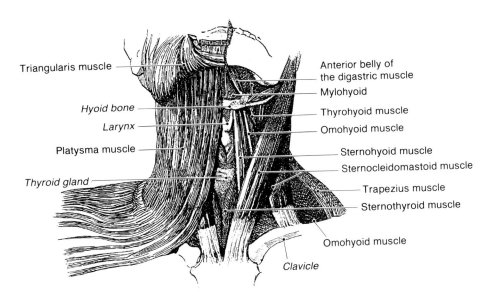

Triangularis muscle

Anterior belly of the digastric muscle

Mylohyoid

Hyoid bone

Thyrohyoid muscle

Larynx

Omohyoid muscle

Platysma muscle

Sternohyoid muscle

Sternocleidomastoid muscle

Thyroid gland

Trapezius muscle

Sternothyroid muscle

Omohyoid muscle

Clavicle

**Figure 131**  *Muscles of the head and neck. As appeared in* An Atlas of Anatomy for Artists. *By Fritz Schider. Dover Publications, Inc. NY.*

# HEAD

| Name | Action | Origin | Insertion |
|------|--------|--------|-----------|
| CORRUGATOR | Draws eyebrows toward midline | Tissue above eyebrow | Root of nose |
| LEVATOR PALPEBRAE | Opens upper eyelid | Orbital roof | Skin of upper eyelid |
| ORBICULARIS OCULI | Firmly closes eyelid | Rim of orbit | Edge of lidslit |
| EPICRANIUS: <br> (a) Occipital belly <br> (b) Frontal belly | (a) Moves scalp backwards <br> (b) Wrinkles forehead | (a) Occipital bone and mastoid process <br> (b) Sheath of occipital belly | (a) Sheath of frontal belly <br> (b) Skin of eyebrows |
| PROCERUS | Wrinkles skin horizontally at bridge of nose | Nasal bones | Skin in between eyebrows |
| NASALIS | Compresses nostrils | Maxilla & nasal cartilage | Sheath covering nostrils |
| ZYGOMATICUS | Moves corners of mouth outward and upward | Zygomatic bone | ORBICULARIS ORIS and skin at corners of mouth |
| MASSETER | Raises lower jaw for chewing | Zygomatic arch | Mandible |
| RISORIUS | Dimples corners of mouth | Hollow of cheek | Orbicularis oris at corner of mouth |
| BUCCINATOR | Compresses cheek | Maxilla | ORBICULARIS ORIS |
| ORBICULARIS ORIS | Protrudes lips; closes and points mouth | Muscles nearby, notably buccinator | Lip-rim |
| DEPRESSOR ANGULIS ORIS | Brings corners of mouth downward | Mandible | Corner of mouth |
| DEPRESSOR LABII INFERIORIS | Moves lower lip downward and outward | Mandible | ORBICULARIS ORIS |
| MENTALIS | Elevates and protrudes lower lip | Mandible (chin) | ORBICULARIS ORIS |
| TEMPORALIS | Raises lower jaw for chewing and speaking | Temporal bone | Mandible |
| STERNOCLEIDOMASTOID | Lifts head; tips head backward; turns head to side | Sternum & clavicle | Mastoid process of the temporal bone of skull |
| TRAPEZIUS | Brings scapulae together; moves them up and down; draws head backwards | Occipital bone (on skull) down to thoracic vertebra #12 | Clavicle, acromion, & spine of the scapula |
| LEVATOR LABII SUPERIORIS | Elevates upper lip | Maxilla & zygomatic bone | ORBICULARIS ORIS & skin above lips |

All actions and descriptive terms are defined with reference to the
ANATOMICAL POSITION, that is, a figure standing erect, feet
parallel and flat on the floor, eyes directed forward, arms at the sides,
and palms facing forward.

## Muscles of Facial Expression

On the following three pages the primary muscles used for each expression have been listed. Place a piece of tracing paper over the photographs and sketch in the appropriate muscles. For your drawing reference, muscle identification is available in the schematic illustration. Use a 2B drawing pencil, or code again choosing a group of colored pencils for consistency in drawing the facial muscles. ▲

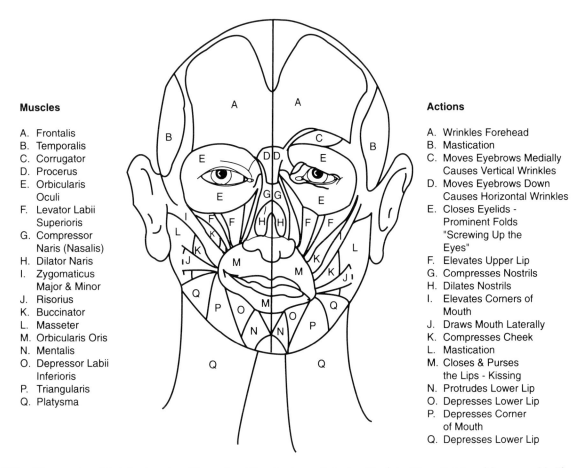

**Muscles**

A. Frontalis
B. Temporalis
C. Corrugator
D. Procerus
E. Orbicularis Oculi
F. Levator Labii Superioris
G. Compressor Naris (Nasalis)
H. Dilator Naris
I. Zygomaticus Major & Minor
J. Risorius
K. Buccinator
L. Masseter
M. Orbicularis Oris
N. Mentalis
O. Depressor Labii Inferioris
P. Triangularis
Q. Platysma

**Actions**

A. Wrinkles Forehead
B. Mastication
C. Moves Eyebrows Medially Causes Vertical Wrinkles
D. Moves Eyebrows Down Causes Horizontal Wrinkles
E. Closes Eyelids - Prominent Folds "Screwing Up the Eyes"
F. Elevates Upper Lip
G. Compresses Nostrils
H. Dilates Nostrils
I. Elevates Corners of Mouth
J. Draws Mouth Laterally
K. Compresses Cheek
L. Mastication
M. Closes & Purses the Lips - Kissing
N. Protrudes Lower Lip
O. Depresses Lower Lip
P. Depresses Corner of Mouth
Q. Depresses Lower Lip

***Figure 132*** *The muscles of facial expression. Figure from* The Anatomy Coloring Book *by Wynn Kapit and Lawrence M. Elson. Copyright © 1977 by Wynn Kapit and Lawrence M. Elson. Reprinted by permission of Harper Collins Publishers.*

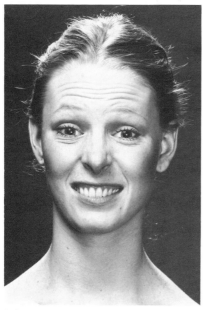

(a.)

A. Frontalis
I. Zygomaticus Major & Minor
H. Dilator Naris
F. Levator labii superioris

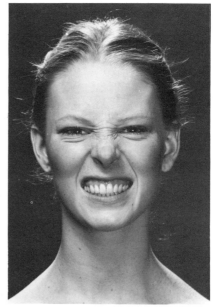

(b.)

D. Procerus
F. Levator labii superioris
H. Dilator Naris
I. Zygomaticus Major & Minor

(c.)

C. Corrugator
E. Orbicularis oculi
K. Buccinator
M. Obicularis oris
G. Compressor naris

(d.)

C. Corrugator
H. Dilator naris
M. Obicularis oris
J. Risorius

*Figure 133a–h*  *Muscles of facial expressions. Courtesy © Dr. Sheril B. Burton.*

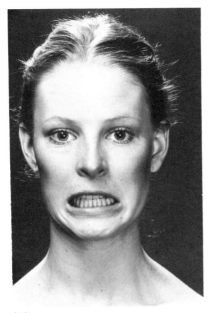

(e.)

M. Obicularis oris
O. Depressor labii inferioris
P. Triangularis
Q. Platysma

(f.)

H. Dilator naris
O. Depressor labii inferioris
N. Mentalis
M. Orbicularis oris
P. Triangularis

(g.)

O. Depressor labii inferioris
P. Triangularis
M. Orbicularis oris
C. Corrugator

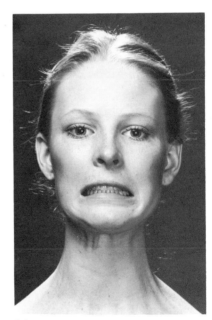

(h.)

O. Depressor labii inferioris
Q. Platysma

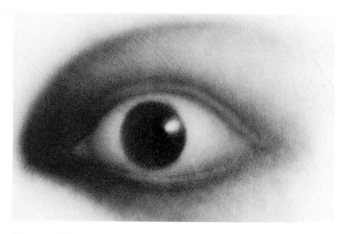

**Figure 134** *Rudolfo Abularach (1933–0000).* Circe. *Ink on paper 23 7/8" × 29". 1969. From San Francisco Museum of Art; Gift of San Francisco Women Artists.*

**Figure 135a–i** *Position of eyes. Courtesy © Dr. Sheril B. Burton.*

(a.)

(b.)

(c.)

(d.)

(e.)

(f.)

(g.)

(h.)

(i.)

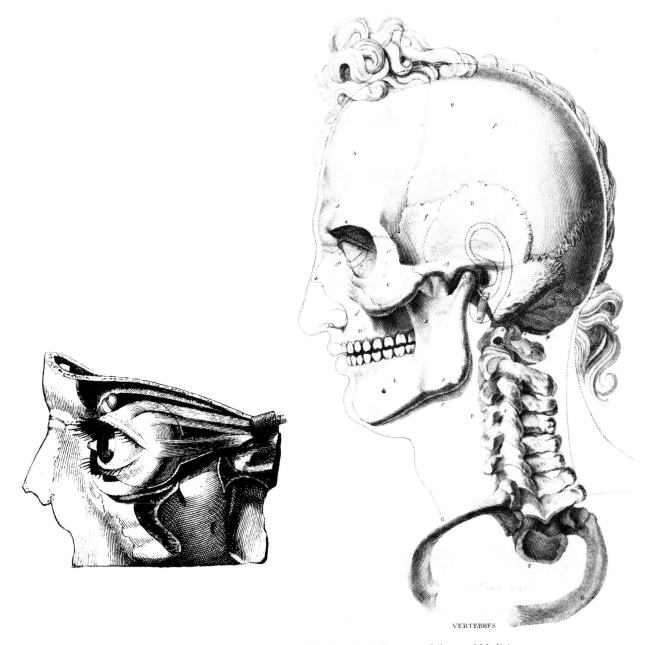

*Figure 136*    *The skull in profile with muscle overlay. Courtesy of the Francis A. Countway Library of Medicine.*

VERTEBRES

**Figure 137**    *Andreas Vesalius (1514–1564). Outermost order of muscles.* The Illustrations from the Works of Andreas Vesalius of Brussels. *By J. B. de C. M. Saunders and Charles D. O'Malley. The World Publishing Co., NY. 1950. p. 35.]*

**Figure 138**    *Andreas Vesalius (1514–1564). Second order of muscles. [The* Illustrations from the Works of Andreas Vesalius of Brussels. *By J. B. de C. M. Saunders and Charles D. O'Malley. The World Publishing Co., NY. 1950. p. 41.]*

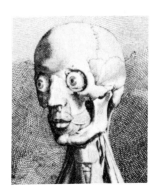

**Figure 139**    *Andreas Vesalius (1514–1564). Third order of muscles. [The* Illustrations from the Works of Andreas Vesalius of Brussels. *By J. B. de C. M. Saunders and Charles D. O'Malley. The World Publishing Co., NY. 1950. p. 45.]*

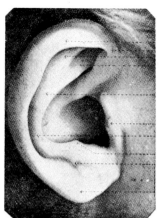

Helix of the ear
Anterior branch of the helix
Superior branch ⎫
Inferior branch ⎬ of the anti-helix

Anti-helix

Tragus
Posterior branch of the anti-helix

Anti-tragus
Lower end of the anti-helix

Lobule of the ear

**Figure 140**   *Surface ear, lateral (side). As appeared in* An Atlas of Anatomy for Artists. *By Fritz Schider. Dover Publications, Inc. NY.*

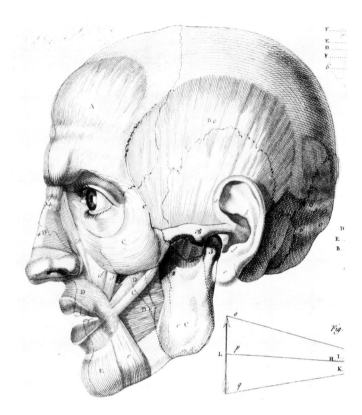

**Figure 141**   *Jean Galbert Savage. Head, lateral (side). [As appeared in* Anatomy of Bones and Muscles Applicable to the Fine Arts. *By Jean Galbert Savage. Courtesy: Boston Medical Library in Francis A. Countway Library of Medicine. Photograph by Kalmon Zabarsky.]*

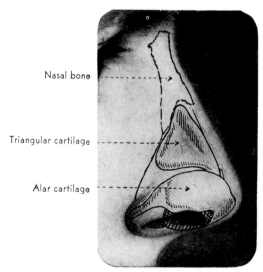

Nasal bone

Triangular cartilage

Alar cartilage

**Figure 142**   *Nose, lateral (side). As appeared in* An Atlas of Anatomy for Artists. *By Fritz Schider. Dover Publications, Inc. NY.*

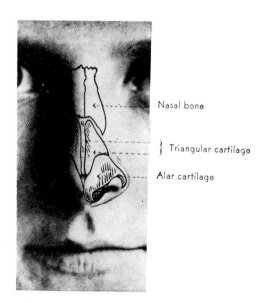

Nasal bone

⎫ Triangular cartilage

Alar cartilage

**Figure 143**   *Nose, anterior (front). As appeared in* An Atlas of Anatomy for Artists. *By Fritz Schider. Dover Publications, Inc. NY.*

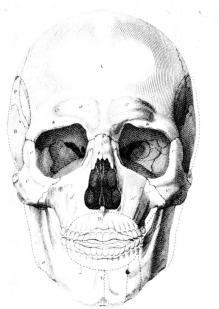

**Figure 144**   *Frontal view of the skull. Courtesy of the Francis A. Countway Library of Medicine.*

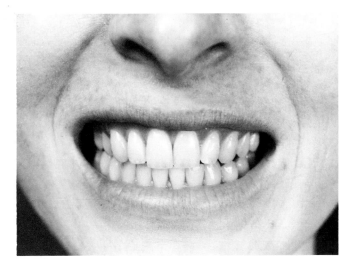

**Figure 145**   *Lips and teeth. Courtesy © Sheril B. Burton.*

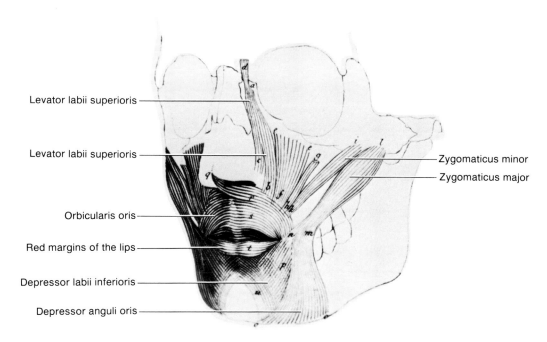

Levator labii superioris

Levator labii superioris

Orbicularis oris

Red margins of the lips

Depressor labii inferioris

Depressor anguli oris

Zygomaticus minor

Zygomaticus major

**Figure 146**   *Albinus (1697–1770). Muscles of the lips and cheek.* [Albinus on Anatomy. *By Robert Beverly Hale and Terrance Coyle. Watson-Guptill Pub., NY. 1979. p. 63, upper left images.*]

The illustration below is the suggested exercise you might use consistently within this muscle section of studying anatomy. Remember, gesture/sighting first, then volumes, then draw the outer contours of the model posing for you. Once those three steps are taken, refer to the muscles from a chart, the book, or a splayed plaster model, and incorporate or superimpose those over your other drawing, either on the original drawing or with a tracing sheet overlay. Bear in mind the origins and insertions of the muscles as well as their names. If reciting names is too awkward as you draw, then move through the memory information before you begin to draw. You will be surprised how much you retain through repetition. Muscles, their origins, insertions, and actions will soon become second nature to you. ▲

**Figure 147**  *The model's head with volumes, sighting, and features.*

**Figure 148**  *The muscles drawn on the tracing overlay sheet.*

**Figure 149** *The head with muscles superimposed.*

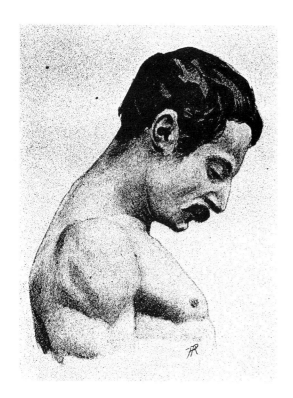

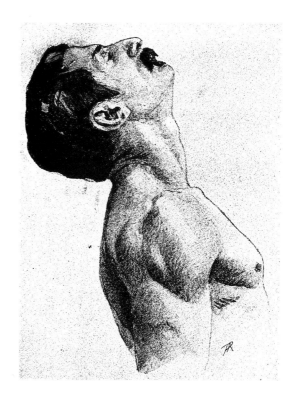

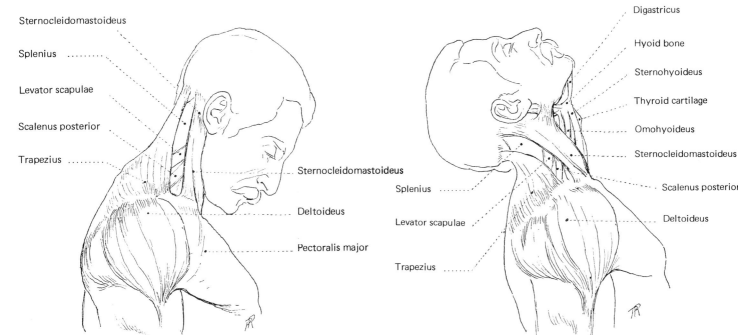

Sternocleidomastoideus

Splenius .........

Levator scapulae

Scalenus posterior

Trapezius .........

Sternocleidomastoideus

Deltoideus

Pectoralis major

Digastricus

Hyoid bone

Sternohyoideus

Thyroid cartilage

Omohyoideus

Sternocleidomastoideus

Scalenus posterior

Deltoideus

Splenius .........

Levator scapulae

Trapezius .........

**Figure 150**   Paul Richer. *Flexion movements of the head and neck.* [Artistic Anatomy. *By Dr. Paul Richer. Translated and edited by Robert Beverly Hale. Watson-Guptill Publ., NY. 1986. pl. 87.*]

**Figure 151**   Paul Richer. *Extension movements of the head and neck.* [Artistic Anatomy. *By Dr. Paul Richer. Translated by Robert Beverly Hale. Watson-Guptill Publ., NY. 1986. pl. 87.*]

138

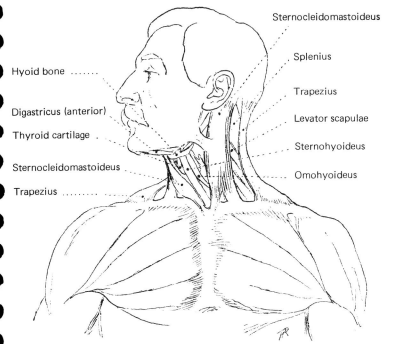

Hyoid bone .......

Digastricus (anterior)

Thyroid cartilage .

Sternocleidomastoideus .........

Trapezius .........

Sternocleidomastoideus

Splenius

Trapezius

Levator scapulae

Sternohyoideus

Omohyoideus

**Figure 152** *Paul Richer. Rotation movements of the head and neck. [Artistic Anatomy. By Dr. Paul Richer. Translated and edited by Robert Beverly Hale. Watson-Guptill Publ., NY. 1986. pl. 88.]*

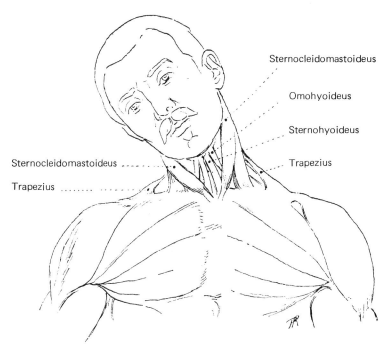

Sternocleidomastoideus .........

Trapezius .......

Sternocleidomastoideus

Omohyoideus

Sternohyoideus

Trapezius

**Figure 153** *Paul Richer. Lateral inclination movements of the head and neck. [Artistic Anatomy. By Dr. Paul Richer. Translated and edited by Robert Beverly Hale. Watson-Guptill Publ., NY. 1986. pl. 88.]*

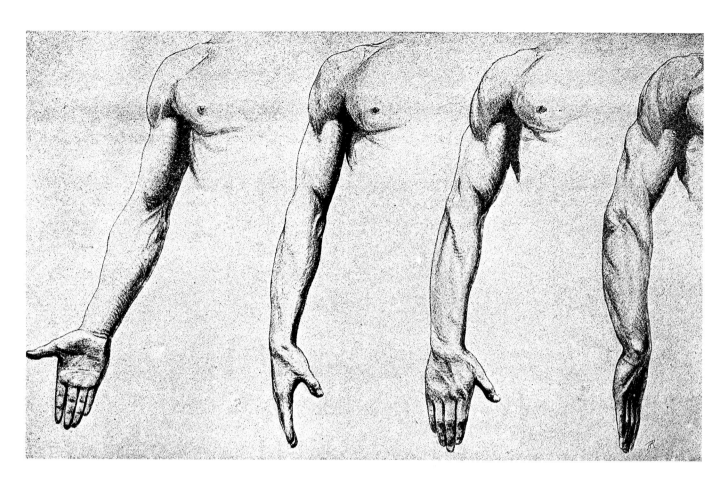

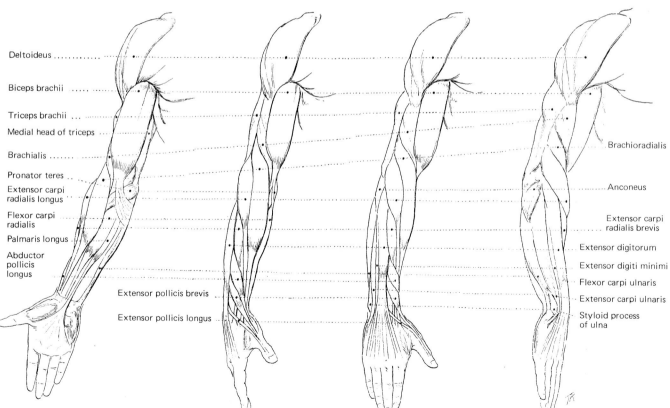

Deltoideus . . . . . . . . . . .

Biceps brachii . . . . .

Triceps brachii . . .

Medial head of triceps

Brachialis . . . . . .

Pronator teres . . .

Extensor carpi
radialis longus

Flexor carpi
radialis

Palmaris longus

Abductor
pollicis
longus

Extensor pollicis brevis

Extensor pollicis longus

Brachioradialis

Anconeus

Extensor carpi
radialis brevis

Extensor digitorum

Extensor digiti minimi

Flexor carpi ulnaris

Extensor carpi ulnaris

Styloid process
of ulna

**Figure 154** *Paul Richer. Movements of the upper limb.* [Artistic Anatomy. By Dr. Paul Richer. Translated and edited by Robert Beverly Hale. Watson-Guptill Publ., NY. 1986. pl. 101.]

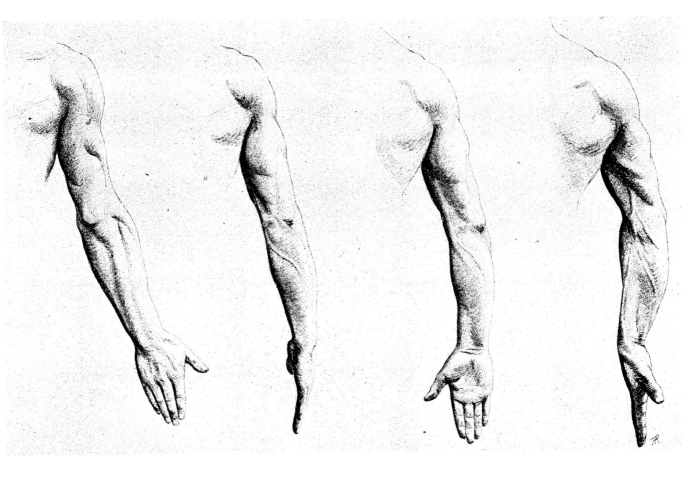

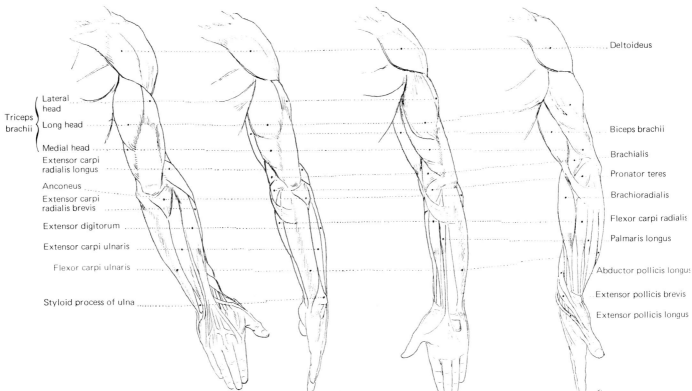

Triceps brachii
{
Lateral head
Long head
Medial head
}

Extensor carpi radialis longus

Anconeus

Extensor carpi radialis brevis

Extensor digitorum

Extensor carpi ulnaris

Flexor carpi ulnaris

Styloid process of ulna

Deltoideus

Biceps brachii

Brachialis

Pronator teres

Brachioradialis

Flexor carpi radialis

Palmaris longus

Abductor pollicis longus

Extensor pollicis brevis

Extensor pollicis longus

**Figure 155**   *Paul Richer. Movements of the upper limb. [Artistic Anatomy. By Dr. Paul Richer. Translated and edited by Robert Beverly Hale. Watson-Guptill Publ., NY. 1986. pl. 102.]*

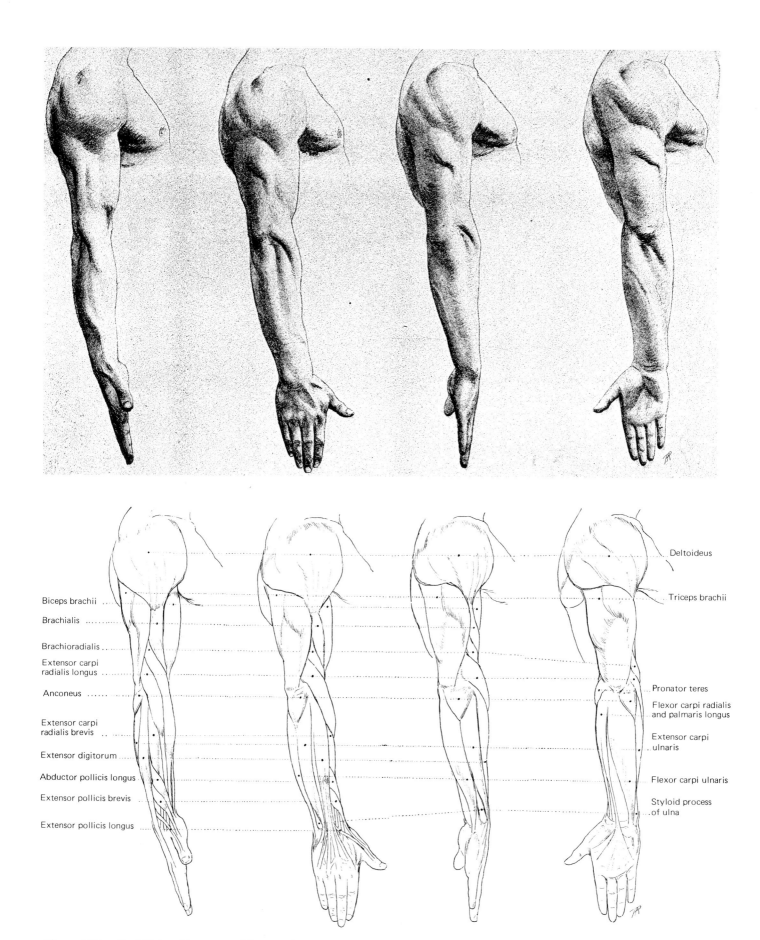

Biceps brachii

Brachialis

Brachioradialis

Extensor carpi
radialis longus

Anconeus

Extensor carpi
radialis brevis

Extensor digitorum

Abductor pollicis longus

Extensor pollicis brevis

Extensor pollicis longus

Deltoideus

Triceps brachii

Pronator teres

Flexor carpi radialis
and palmaris longus

Extensor carpi
ulnaris

Flexor carpi ulnaris

Styloid process
of ulna

***Figure 156*** *Paul Richer. Movements of the upper limb. [Artistic Anatomy. By Dr. Paul Richer. Translated and edited by Robert Beverly Hale. Watson-Guptill Publ., NY. 1986. pl. 103.]*

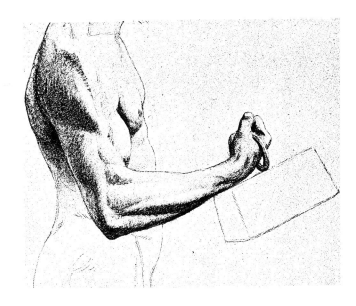

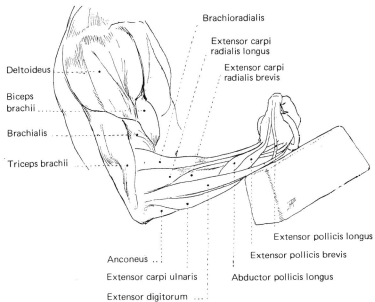

Brachioradialis

Extensor carpi
radialis longus

Extensor carpi
radialis brevis

Deltoideus

Biceps
brachii

Brachialis

Triceps brachii

Extensor pollicis longus

Anconeus ...

Extensor pollicis brevis

Extensor carpi ulnaris

Abductor pollicis longus

Extensor digitorum ...

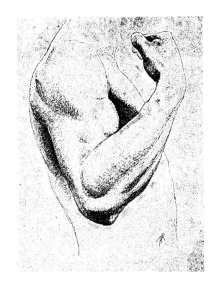

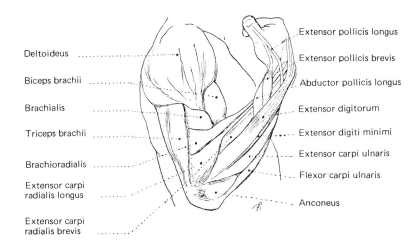

Deltoideus

Biceps brachii

Brachialis

Triceps brachii

Brachioradialis

Extensor carpi
radialis longus

Extensor carpi
radialis brevis

Extensor pollicis longus

Extensor pollicis brevis

Abductor pollicis longus

Extensor digitorum

Extensor digiti minimi

Extensor carpi ulnaris

Flexor carpi ulnaris

Anconeus

**Figure 157** *Paul Richer. Movements of the upper limb.* [Artistic Anatomy. *By Dr. Paul Richer. Translated and edited by Robert Beverly Hale. Watson-Guptill Publ., NY. 1986. pl. 104.*]

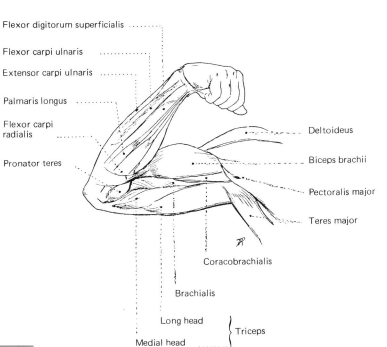

Flexor digitorum superficialis

Flexor carpi ulnaris

Extensor carpi ulnaris

Palmaris longus

Flexor carpi
radialis

Pronator teres

Deltoideus

Biceps brachii

Pectoralis major

Teres major

Coracobrachialis

Brachialis

Long head

Medial head

} Triceps

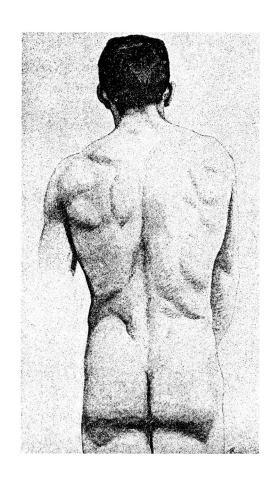

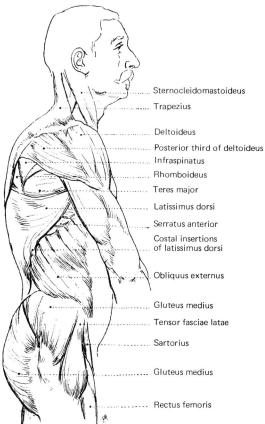

Sternocleidomastoideus

Trapezius

Deltoideus

Posterior third of deltoideus

Infraspinatus

Rhomboideus

Teres major

Latissimus dorsi

Serratus anterior

Costal insertions
of latissimus dorsi

Obliquus externus

Gluteus medius

Tensor fasciae latae

Sartorius

Gluteus medius

Rectus femoris

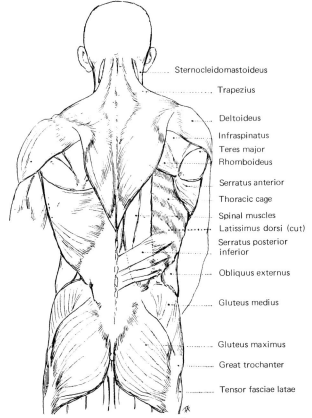

Sternocleidomastoideus

Trapezius

Deltoideus

Infraspinatus

Teres major

Rhomboideus

Serratus anterior

Thoracic cage

Spinal muscles

Latissimus dorsi (cut)

Serratus posterior
inferior

Obliquus externus

Gluteus medius

Gluteus maximus

Great trochanter

Tensor fasciae latae

**Figure 158**   *Paul Richer. Movements of the shoulder. [Artistic Anatomy. By Dr. Paul Richer. Translated and edited by Robert Beverly Hale. Watson-Guptill Publ., NY. 1986. pl. 89.]*

**Figure 159**   *Paul Richer. Movements of the shoulder. [Artistic Anatomy. By Dr. Paul Richer. Translated and edited by Robert Beverly Hale. Watson-Guptill Publ., NY. 1986. pl. 90.]*

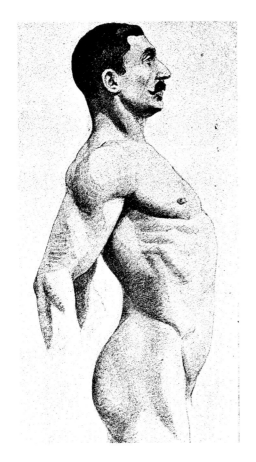

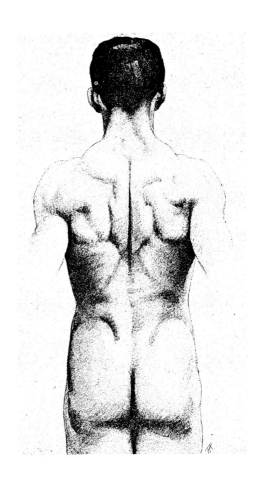

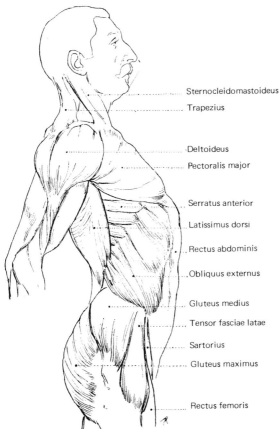

Sternocleidomastoideus

Trapezius

Deltoideus

Pectoralis major

Serratus anterior

Latissimus dorsi

Rectus abdominis

Obliquus externus

Gluteus medius

Tensor fasciae latae

Sartorius

Gluteus maximus

Rectus femoris

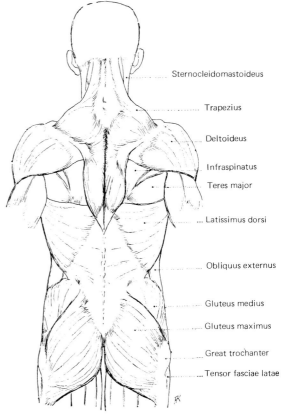

Sternocleidomastoideus

Trapezius

Deltoideus

Infraspinatus

Teres major

Latissimus dorsi

Obliquus externus

Gluteus medius

Gluteus maximus

Great trochanter

Tensor fasciae latae

**Figure 160** *Paul Richer. Movements of the shoulder. [Artistic Anatomy. By Dr. Paul Richer. Translated and edited by Robert Beverly Hale. Watson-Guptill Publ., NY. 1986. pl. 89.]*

**Figure 161** *Paul Richer. Movements of the shoulder. [Artistic Anatomy. By Dr. Paul Richer. Translated and edited by Robert Beverly Hale. Watson-Guptill Publ., NY. 1986. pl. 89.]*

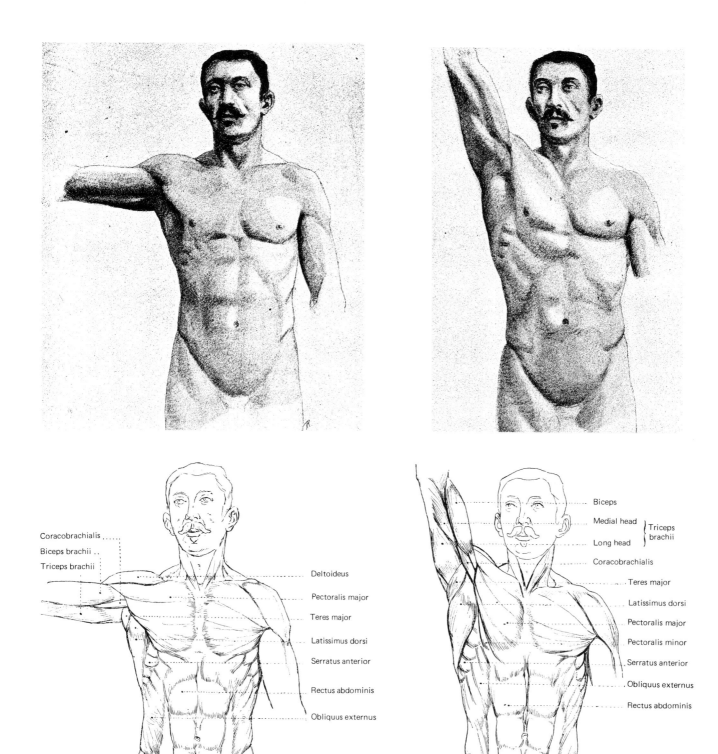

Coracobrachialis
Biceps brachii
Triceps brachii

Deltoideus

Pectoralis major

Teres major

Latissimus dorsi

Serratus anterior

Rectus abdominis

Obliquus externus

Gluteus medius

Tensor fasciae latae

Sartorius

Rectus femoris

Biceps

Medial head } Triceps
                brachii
Long head

Coracobrachialis

Teres major

Latissimus dorsi

Pectoralis major

Pectoralis minor

Serratus anterior

Obliquus externus

Rectus abdominis

Gluteus medius

Tensor fasciae latae

Sartorius

**Figure 162** *Paul Richer. Modifications of exterior form during movements of the arm. [Artistic Anatomy. By Dr. Paul Richer. Translated and edited by Robert Beverly Hale. Watson-Guptill Publ., NY. 1986. pl. 91.]*

**Figure 163** *Paul Richer. Modifications of exterior form during movements of the arm. [Artistic Anatomy. By Dr. Paul Richer. Translated and edited by Robert Beverly Hale. Watson-Guptill Publ., NY. 1986. pl. 91.]*

146

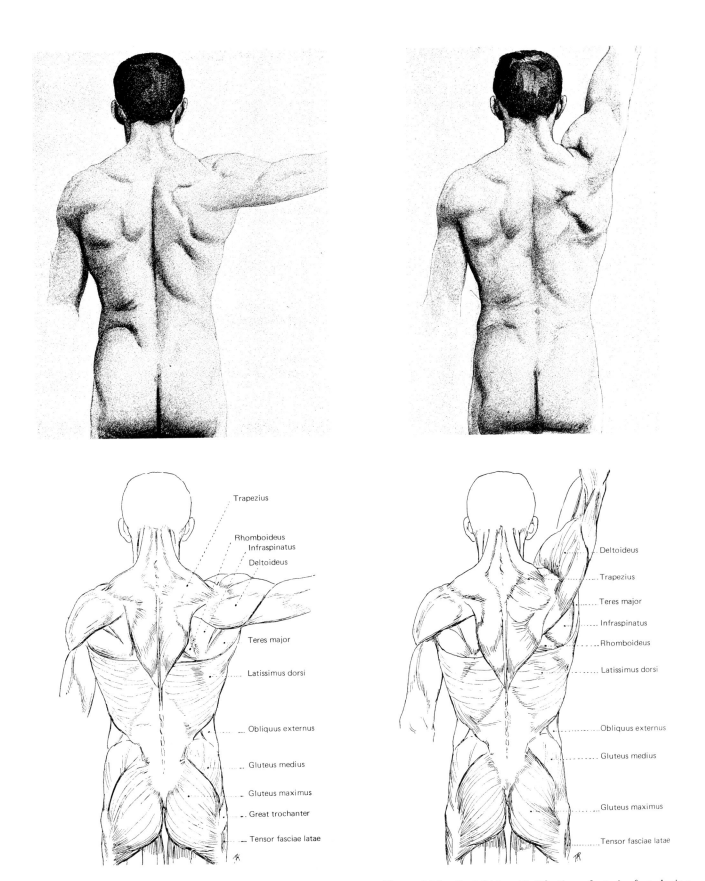

**Figure 164**   *Paul Richer. Modifications of exterior form during movements of the arm. [Artistic Anatomy. By Dr. Paul Richer. Translated and edited by Robert Beverly Hale. Watson-Guptill Publ., NY. 1986. pl. 92.]*

**Figure 165**   *Paul Richer. Modifications of exterior form during movements of the arm. [Artistic Anatomy. By Dr. Paul Richer. Translated and edited by Robert Beverly Hale. Watson-Guptill Publ., NY. 1986. pl. 92.]*

147

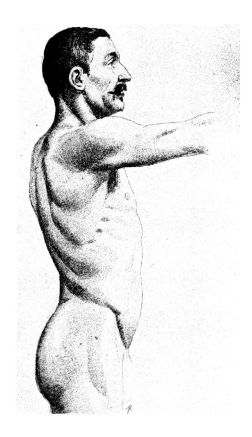

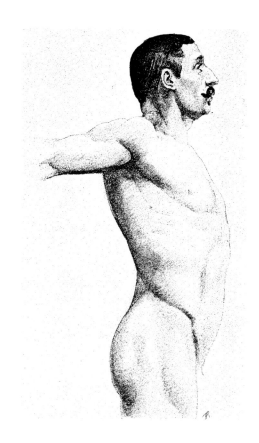

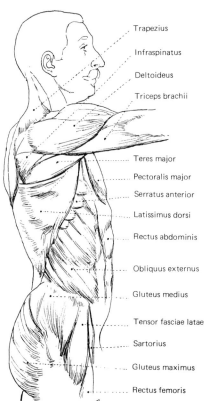

Trapezius

Infraspinatus

Deltoideus

Triceps brachii

Teres major

Pectoralis major

Serratus anterior

Latissimus dorsi

Rectus abdominis

Obliquus externus

Gluteus medius

Tensor fasciae latae

Sartorius

Gluteus maximus

Rectus femoris

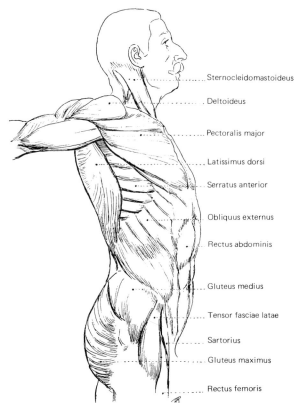

Sternocleidomastoideus

Deltoideus

Pectoralis major

Latissimus dorsi

Serratus anterior

Obliquus externus

Rectus abdominis

Gluteus medius

Tensor fasciae latae

Sartorius

Gluteus maximus

Rectus femoris

***Figure 166*** *Paul Richer. Modifications of exterior form during movements of the arm. [Artistic Anatomy. By Dr. Paul Richer. Translated and edited by Robert Beverly Hale. Watson-Guptill Publ., NY. 1986. pl. 93.]*

***Figure 167*** *Paul Richer. Modifications of exterior form during movements of the arm. [Artistic Anatomy. By Dr. Paul Richer. Translated and edited by Robert Beverly Hale. Watson-Guptill Publ., NY. 1986. pl. 93.]*

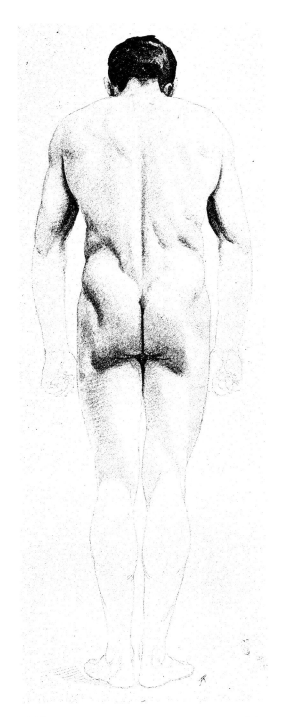
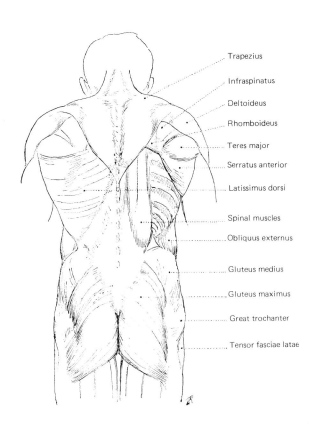

Trapezius

Infraspinatus

Deltoideus

Rhomboideus

Teres major

Serratus anterior

Latissimus dorsi

Spinal muscles

Obliquus externus

Gluteus medius

Gluteus maximus

Great trochanter

Tensor fasciae latae

***Figure 168*** *Paul Richer. Light flexion movements of the trunk. [Artistic Anatomy. By Dr. Paul Richer. Translated and edited by Robert Beverly Hale. Watson-Guptill Publ., NY. 1986. pl. 94.]*

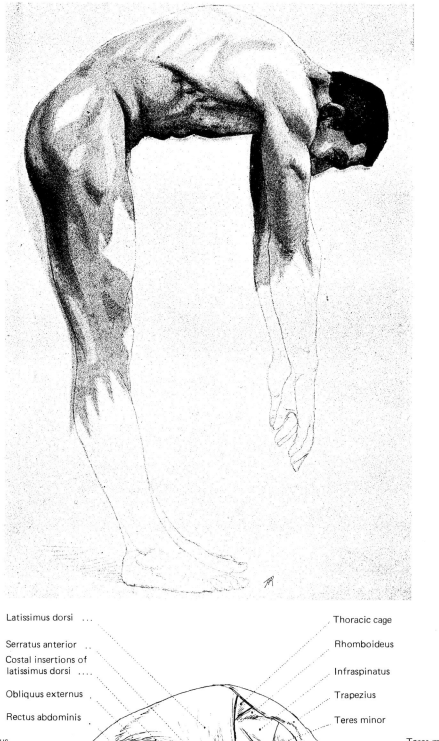

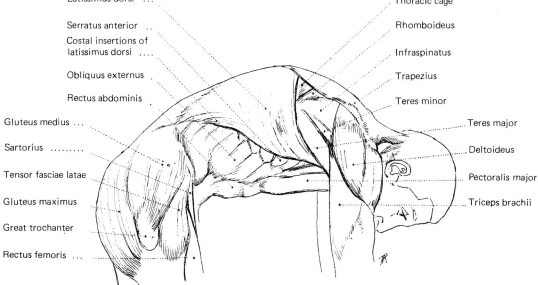

Latissimus dorsi ...                                    Thoracic cage

Serratus anterior ..                                    Rhomboideus

Costal insertions of
latissimus dorsi ....                                   Infraspinatus

Obliquus externus .                                     Trapezius

Rectus abdominis                                        Teres minor

Gluteus medius ...                                      Teres major

Sartorius .........                                     Deltoideus

Tensor fasciae latae                                    Pectoralis major

Gluteus maximus                                         Triceps brachii

Great trochanter

Rectus femoris ...

***Figure 169*** *Paul Richer. Forced flexion movements of the trunk. [Artistic Anatomy. By Dr. Paul Richer. Translated and edited by Robert Beverly Hale. Watson-Guptill Publ., NY. 1986. pl. 95.]*

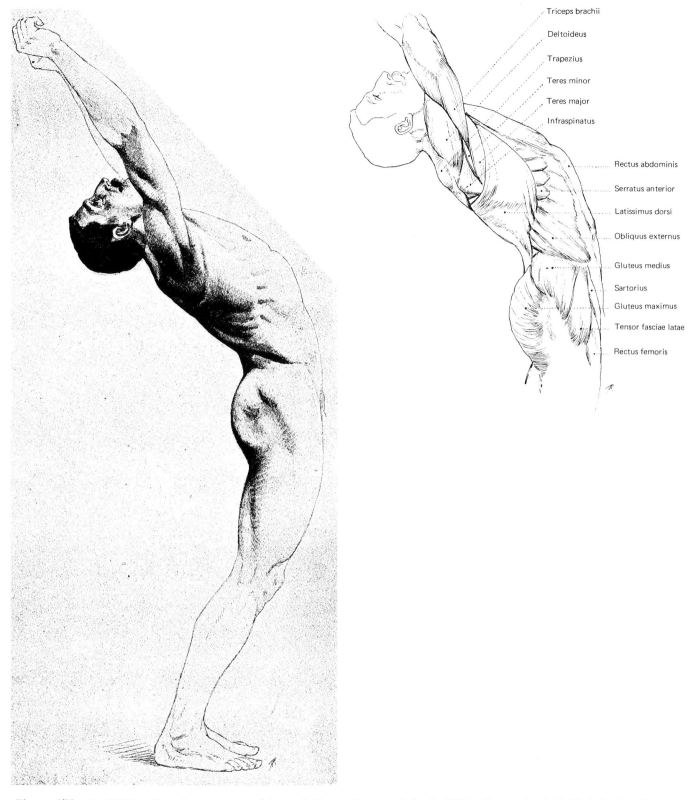

Triceps brachii
Deltoideus
Trapezius
Teres minor
Teres major
Infraspinatus

Rectus abdominis
Serratus anterior
Latissimus dorsi
Obliquus externus
Gluteus medius
Sartorius
Gluteus maximus
Tensor fasciae latae
Rectus femoris

***Figure 170*** *Paul Richer. Extension movements of the trunk. [Artistic Anatomy. By Dr. Paul Richer. Translated and edited by Robert Beverly Hale. Watson-Guptill Publ., NY. 1986. pl. 96.]*

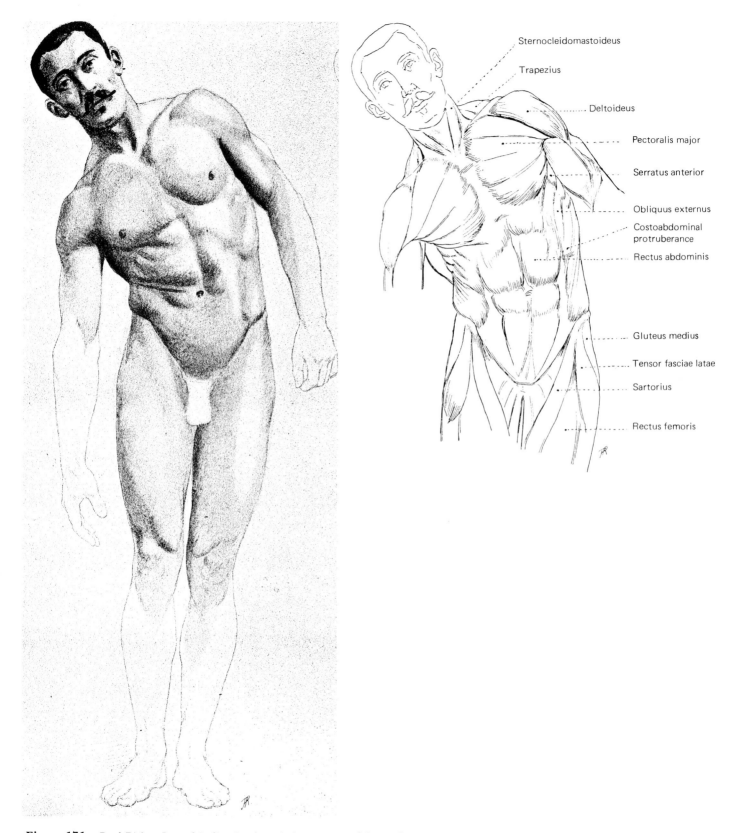

Sternocleidomastoideus

Trapezius

Deltoideus

Pectoralis major

Serratus anterior

Obliquus externus

Costoabdominal
protruberance

Rectus abdominis

Gluteus medius

Tensor fasciae latae

Sartorius

Rectus femoris

***Figure 171*** *Paul Richer. Lateral inclination (anterior) movements of the trunk. [*Artistic Anatomy. By Dr. Paul Richer. Translated and edited by
Robert Beverly Hale. Watson-Guptill Publ., NY. 1986. pl. 97.]*

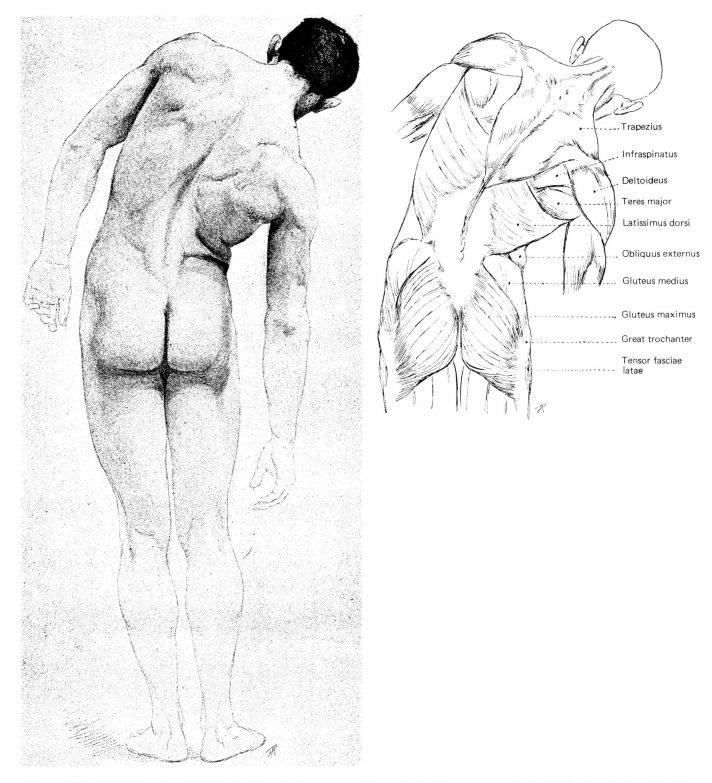

Trapezius

Infraspinatus

Deltoideus

Teres major

Latissimus dorsi

Obliquus externus

Gluteus medius

Gluteus maximus

Great trochanter

Tensor fasciae latae

**Figure 172**   *Paul Richer. Lateral inclination (posterior) movements of the trunk. [*Artistic Anatomy. *By Dr. Paul Richer. Translated and edited by Robert Beverly Hale. Watson-Guptill Publ., NY. 1986. pl. 98.]*

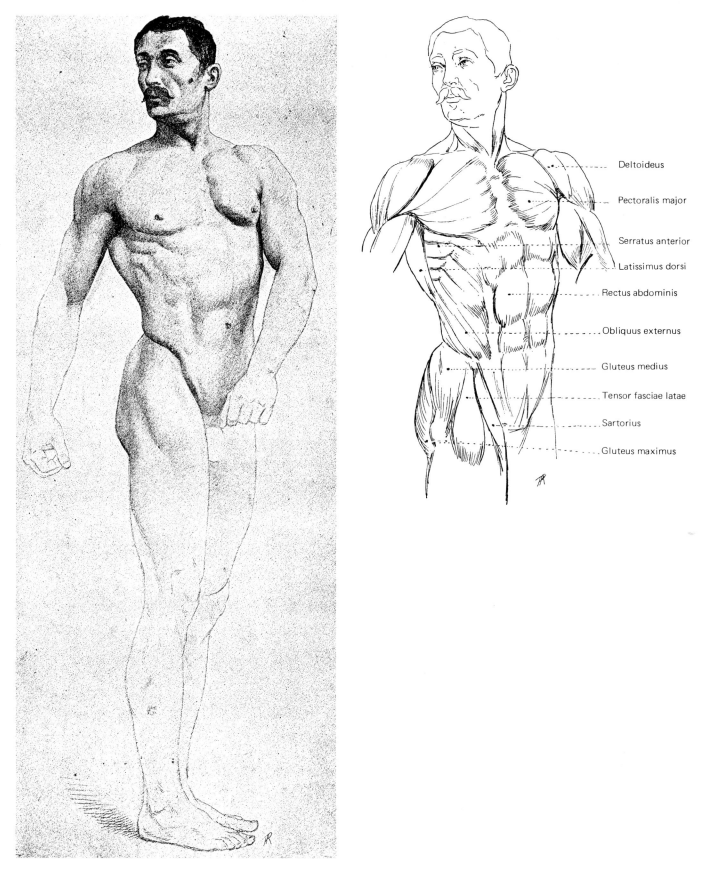

Deltoideus

Pectoralis major

Serratus anterior

Latissimus dorsi

Rectus abdominis

Obliquus externus

Gluteus medius

Tensor fasciae latae

Sartorius

Gluteus maximus

**Figure 173**   *Paul Richer. Rotation (to the right) movements of the trunk. [*Artistic Anatomy. *By Dr. Paul Richer. Translated and edited by Robert Beverly Hale. Watson-Guptill Publ., NY. 1986. pl. 99.]*

154

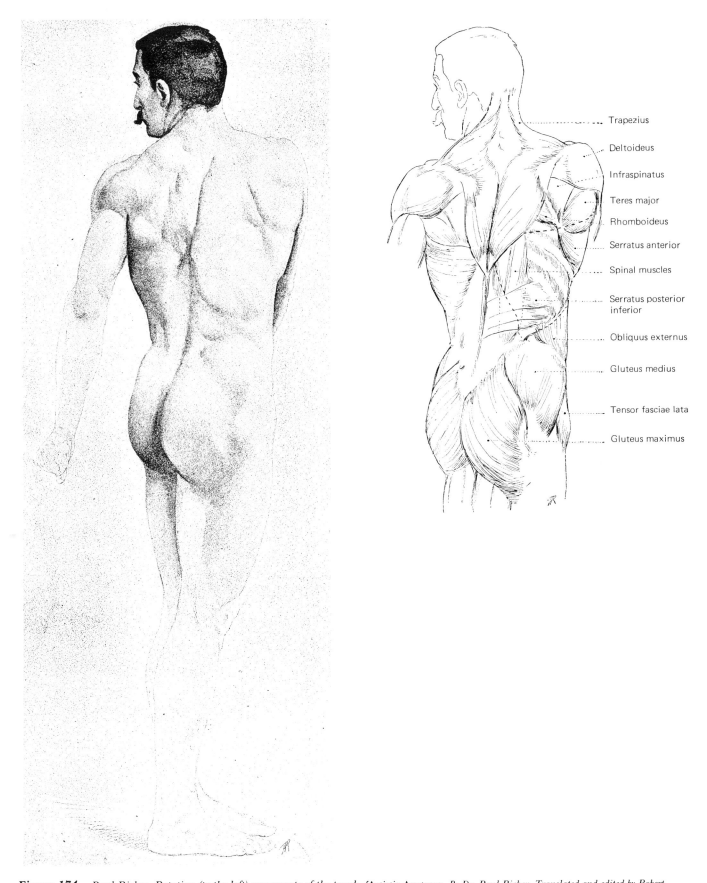

Trapezius

Deltoideus

Infraspinatus

Teres major

Rhomboideus

Serratus anterior

Spinal muscles

Serratus posterior
inferior

Obliquus externus

Gluteus medius

Tensor fasciae lata

Gluteus maximus

**Figure 174** *Paul Richer. Rotation (to the left) movements of the trunk. [Artistic Anatomy. By Dr. Paul Richer. Translated and edited by Robert Beverly Hale. Watson-Guptill Publ., NY. 1986. pl. 100.]*

155

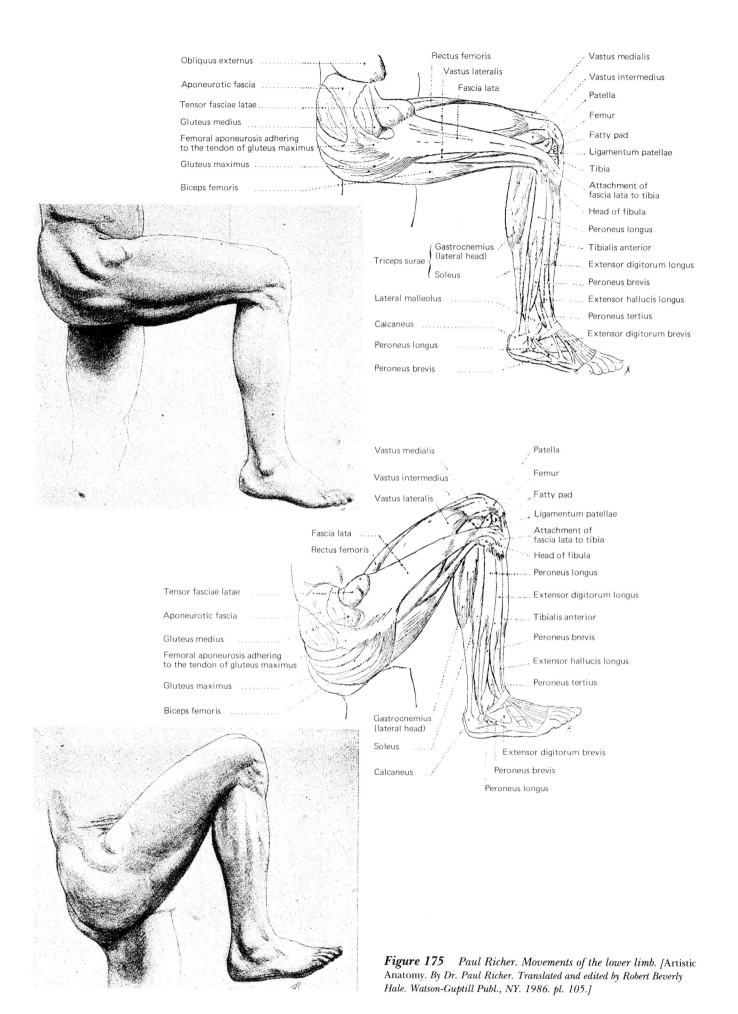

Obliquus externus

Aponeurotic fascia

Tensor fasciae latae

Gluteus medius

Femoral aponeurosis adhering
to the tendon of gluteus maximus

Gluteus maximus

Biceps femoris

Rectus femoris

Vastus lateralis

Fascia lata

Vastus medialis

Vastus intermedius

Patella

Femur

Fatty pad

Ligamentum patellae

Tibia

Attachment of
fascia lata to tibia

Head of fibula

Peroneus longus

Tibialis anterior

Extensor digitorum longus

Peroneus brevis

Extensor hallucis longus

Peroneus tertius

Extensor digitorum brevis

Triceps surae { Gastrocnemius (lateral head) / Soleus

Lateral malleolus

Calcaneus

Peroneus longus

Peroneus brevis

Vastus medialis

Vastus intermedius

Vastus lateralis

Patella

Femur

Fatty pad

Ligamentum patellae

Attachment of
fascia lata to tibia

Head of fibula

Peroneus longus

Extensor digitorum longus

Tibialis anterior

Peroneus brevis

Extensor hallucis longus

Peroneus tertius

Fascia lata

Rectus femoris

Tensor fasciae latae

Aponeurotic fascia

Gluteus medius

Femoral aponeurosis adhering
to the tendon of gluteus maximus

Gluteus maximus

Biceps femoris

Gastrocnemius
(lateral head)

Soleus

Calcaneus

Extensor digitorum brevis

Peroneus brevis

Peroneus longus

***Figure 175*** *Paul Richer. Movements of the lower limb. [Artistic Anatomy. By Dr. Paul Richer. Translated and edited by Robert Beverly Hale. Watson-Guptill Publ., NY. 1986. pl. 105.]*

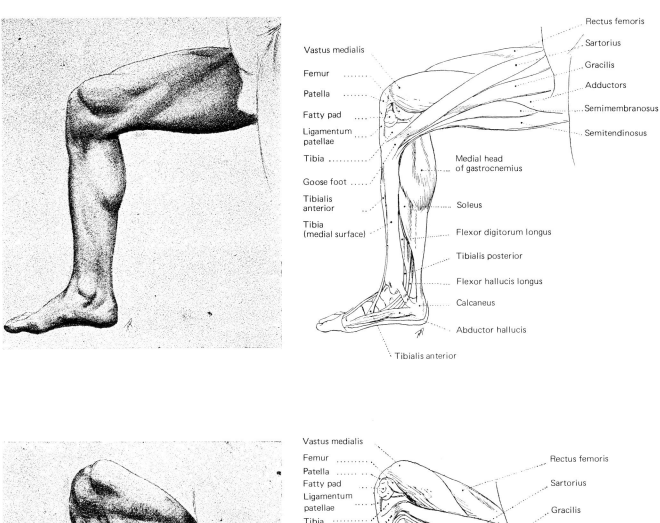

Vastus medialis

Femur

Patella

Fatty pad

Ligamentum
patellae

Tibia

Goose foot

Tibialis
anterior

Tibia
(medial surface)

Medial head
of gastrocnemius

Soleus

Flexor digitorum longus

Tibialis posterior

Flexor hallucis longus

Calcaneus

Abductor hallucis

Tibialis anterior

Rectus femoris

Sartorius

Gracilis

Adductors

Semimembranosus

Semitendinosus

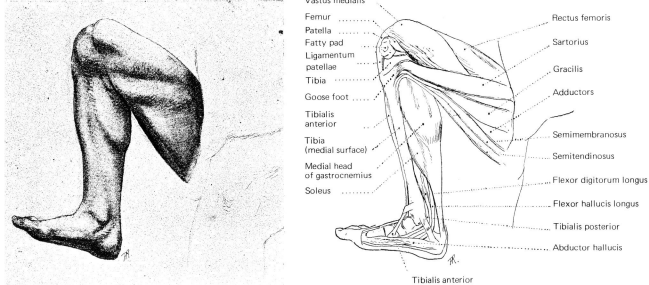

Vastus medialis

Femur

Patella

Fatty pad

Ligamentum
patellae

Tibia

Goose foot

Tibialis
anterior

Tibia
(medial surface)

Medial head
of gastrocnemius

Soleus

Rectus femoris

Sartorius

Gracilis

Adductors

Semimembranosus

Semitendinosus

Flexor digitorum longus

Flexor hallucis longus

Tibialis posterior

Abductor hallucis

Tibialis anterior

***Figure 176*** *Paul Richer. Movements of the lower limb. [*Artistic Anatomy. *By Dr. Paul Richer. Translated and edited by Robert Beverly Hale. Watson-Guptill Publ., NY. 1986. pl. 106.]*

157

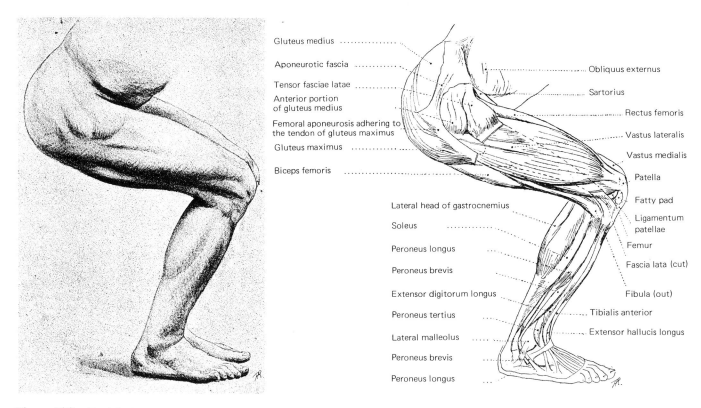

Gluteus medius
Aponeurotic fascia
Tensor fasciae latae
Anterior portion
of gluteus medius
Femoral aponeurosis adhering to
the tendon of gluteus maximus
Gluteus maximus
Biceps femoris

Lateral head of gastrocnemius
Soleus
Peroneus longus
Peroneus brevis
Extensor digitorum longus
Peroneus tertius
Lateral malleolus
Peroneus brevis
Peroneus longus

Obliquus externus
Sartorius
Rectus femoris
Vastus lateralis
Vastus medialis
Patella
Fatty pad
Ligamentum
patellae
Femur
Fascia lata (cut)
Fibula (out)
Tibialis anterior
Extensor hallucis longus

**Figure 177**   *Paul Richer. Movements of the lower limb. [*Artistic Anatomy. *By Dr. Paul Richer. Translated and edited by Robert Beverly Hale. Watson-Guptill Publ., NY. 1986. pl. 107.]*

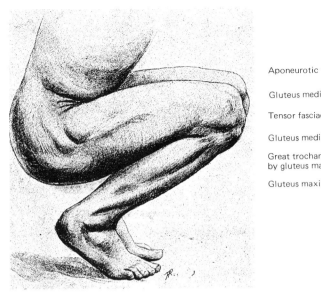

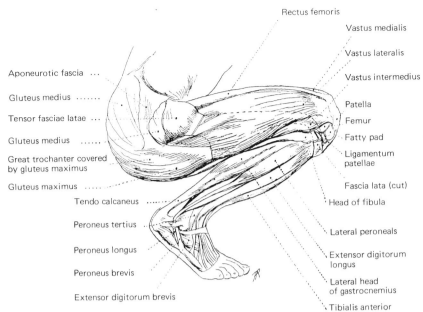

Rectus femoris

Vastus medialis

Vastus lateralis

Vastus intermedius

Aponeurotic fascia ...

Gluteus medius .......

Tensor fasciae latae ...

Gluteus medius ......

Great trochanter covered
by gluteus maximus

Gluteus maximus .....

Patella

Femur

Fatty pad

Ligamentum
patellae

Fascia lata (cut)

Head of fibula

Tendo calcaneus ......

Peroneus tertius ....

Peroneus longus .

Peroneus brevis .

Extensor digitorum brevis

Lateral peroneals

Extensor digitorum
longus

Lateral head
of gastrocnemius

Tibialis anterior

**Figure 177**   (continued)

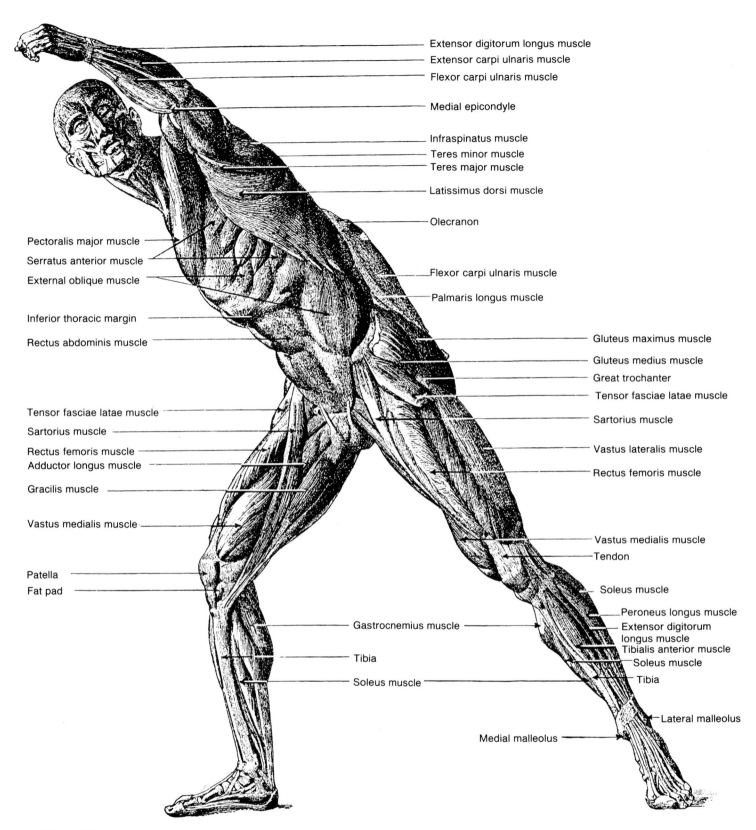

Extensor digitorum longus muscle
Extensor carpi ulnaris muscle
Flexor carpi ulnaris muscle

Medial epicondyle

Infraspinatus muscle
Teres minor muscle
Teres major muscle

Latissimus dorsi muscle

Olecranon

Pectoralis major muscle
Serratus anterior muscle
External oblique muscle

Flexor carpi ulnaris muscle

Palmaris longus muscle

Inferior thoracic margin

Rectus abdominis muscle

Gluteus maximus muscle
Gluteus medius muscle
Great trochanter
Tensor fasciae latae muscle

Tensor fasciae latae muscle
Sartorius muscle

Sartorius muscle

Rectus femoris muscle
Adductor longus muscle

Vastus lateralis muscle
Rectus femoris muscle

Gracilis muscle

Vastus medialis muscle

Vastus medialis muscle
Tendon

Patella
Fat pad

Soleus muscle

Peroneus longus muscle
Extensor digitorum longus muscle
Tibialis anterior muscle
Soleus muscle
Tibia

Gastrocnemius muscle

Tibia

Soleus muscle

Lateral malleolus

Medial malleolus

***Figure 178*** *Muscles in position of* The Fighter *by Borghese. As appeared in* An Atlas of Anatomy for Artists. *By Fritz Schider. Dover Publications, Inc. NY.*

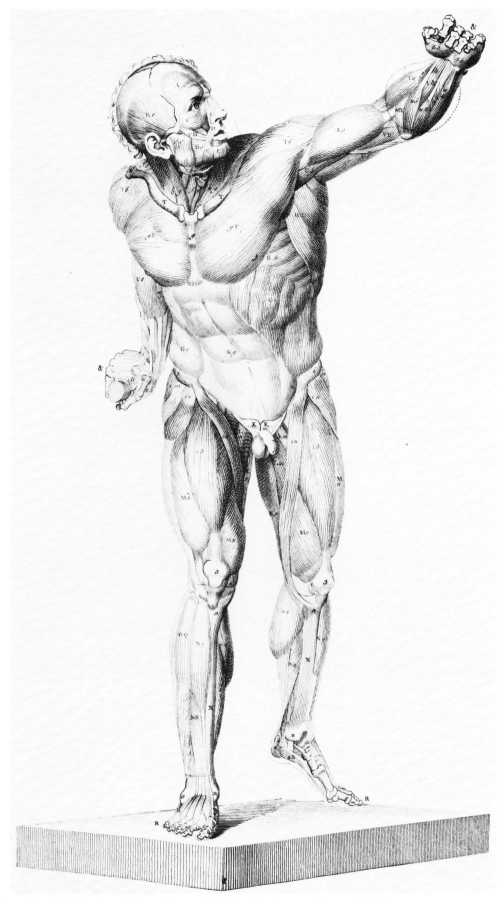

**Figure 179**  *Front view of the Borghese Fighter. Courtesy of the Francis A. Countway Library of Medicine.*

To complete your study of muscles you will need a photograph, preferably one of an active nude figure. If not a nude, then an active figure partially clothed, like a dancer or an athlete. The photograph should be large, at least 8½″ × 11″ and have a sharp focus. Barring a photograph of a live model, your next best selection would be a picture of a statue from the Greek Classical period, the Italian Renaissance, or the Baroque era.

Once found, the subject matter in your picture will need to be enlarged to the scale of your drawing paper, 24″ × 18″. Rescaling can be done fairly simply. If your photo is 11″ × 8½″, and your drawing paper is 24″ × 18″, what scale would you use to enlarge the image to fit on the drawing paper? Selecting one inch squares is a simple example for indicating the process.

Twenty-four divided by eleven is nearly double the longest dimension. And eighteen divided by eight and one half is also nearly double. The graphing scale of 1″ squares on the photo would be 2″ squares on the drawing paper. The ratio would be 1:2. Of course, you could be more precise mathematically. Whatever system works for you, transfer the figure of the photo and its one-inch scale to the paper with its two-inch scale, placing guideline points as you go. Guideline points were referred to in the "sighting gestures" of the skill studies section of *Figure and Form, Vol. I.*

Before drawing the contour of the surface anatomy, use the guideline points as references to draw in gesture/sighting lines very lightly. Move to the volumetric shapes, again, lightly. Now, draw the contours of the body surface. Once completed, you might need to erase back or lighten some of the structural lines, such as the gestural or geometric lines, just enough to retain the new and more complex guidelines as you move to impose bone and muscle.

Try to imagine the figure in the photograph as a live model, anticipating the full move, of which you have a stopframe. Concentrating on a live figure with muscle masses fosters a more informed drawing line. Thinking of the photographic figurative image as a simple flat diagram tends to yield flat, diagrammatic figures. Even though you use one graph over the photo to transpose to a larger graph on the drawing paper, along with guidepoints, try to perceive the figure as though the person were before you. The process is a bit of a contradiction, graphing and thinking masses at the same time, but that mental test also helps line quality.

**Figure 180**  *Stephen Rogers Peck. Figure taken from* Atlas of Human Anatomy for the Artist. *Oxford University Press. 1951. p. 184. fig. 4.*

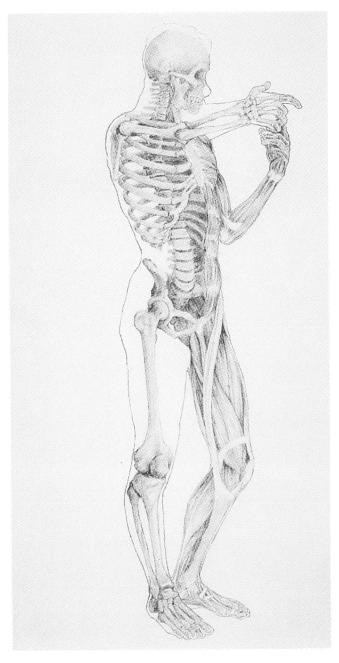

The object of this study is to draw both bones *and* muscles. You might decide to draw two separate drawings, one bones, the other muscles, or both in the same drawing, one side of the figure as bone, the other side as muscle, as in our illustration.

While working your way through the figure's anatomy, access whatever books you need in addition to this textbook. Medical anatomical books may illustrate more views or give more information. Use them if they help.

Drawing paper should be of good quality, preferably 100% rag. The bones and surface contour could be drawn with a 2B pencil. Muscles might be color coded according to an earlier drawing of muscles. Refer back to your anatomical color coding to check for possibilities. By now you understand that using colored pencils can be frustrating when coding very small muscle groups simply because colored leads are soft. A nearby pencil sharpener is almost a necessity.

When you use these drawings as reference points for solving problems in your own work, try to ask or define the following:

1. How was the body unit built? Where did the artist begin?
2. What choices were made about proportioning relative to perspective?
3. Which lines, and what kinds, reveal the most about each unit? Why?
4. By emphasis, attitude, lighting, detail, inference, and expressiveness, each artist is unique. State how you perceive that uniqueness.

**Figure 181** *Michael Thompson. Skeleton and superficial muscles.*
*[Colored pencil, 17 1/2" X 6 3/4". 1988.]*

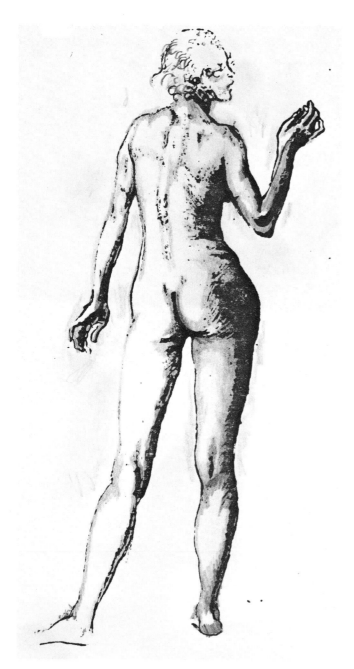

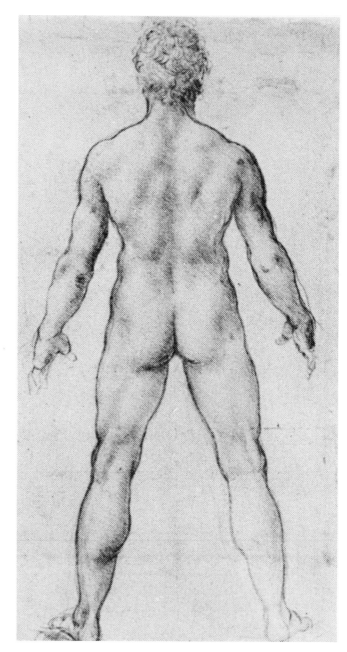

**Figure 182** *Augustus John.* Nude Study. *Pencil, 43.3 × 29.2 cm From Peter Harris Collection.* [Nude to Naked *by Greg Eisler.*]

**Figure 183** *Leonardo da Vinci (1452–1519).* Nude Man Study, back to Spectator. *Red chalk, 10 5/8 × 6 5/16. From Royal Library, Windsor.*

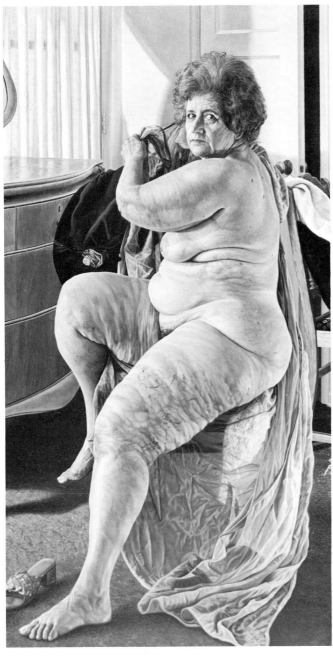

**Figure 184** *James Valerio.* Honey Bun. *Oil on canvas, 183 × 213 cm. From collection of artist. 1974.*

**Figure 185** *Peter Paul Rubens (1577–1640). after Raphael.* Naked Man Dropping from a Wall. *From Payne Knight Bequest, 1824. British Museum. London, England [*Rubens Drawings and Sketches. *p. 41.]*

**Figure 186** *Alfred Leslie.* Julie Schwer and Jane Schwer. *9' × 6'. 1974. Collection of the artist. (Photo: Courtesy Richard Bellamy Gallery.) [Art in America. Jan/Feb 1975. p. 37.]*

**Figure 187** *Michelangelo (1475–1564).* The Risen Christ. *Black chalk 41.4 × 27.4 cm. From British Museum [Drawings of Michelangelo, British Museum, p. 50.]*

(a.)

(b.)

**Figure 188** *(a)* The Artist's Eyes. *[Close Portraits, a catalogue by Lisa Lyons and Martin Friedman. Walker Art Center, Minneapolis, MN. 1980. (b) Detail. Collection of Mrs. Robert B. Mayer, Chicago.]*

*Figure 189* *Andrea de Verrochio (1435–1488).* Head of an Angel. *Picture Gallery, Christ Church, Ox.*

*Figure 190* *Daniel Quintero.* Strange Mongolia. *Sepia on paper, 11 × 28.5 cm. Fine Arts Catalogue. London Arts Council. London, England.*

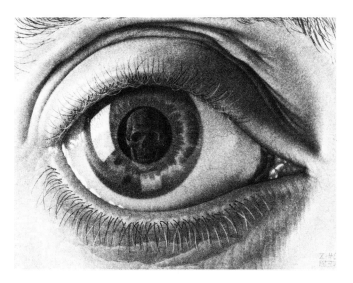

*Figure 191* *M. C. Escher.* Eye. Mezzotint, 6″ × 8″ (151 × 203 cm). X-46 MCE. © 1946 M. C. Escher/Cordon Art-Baarn-Hollard.

*Figure 192* *Antonio Donghi.* The Circus. *Oil on canvas. 1927.* [Art in America. *March 1988. text p. 131.]*

Figure 193 *Jim Dine.* The Potter, Mary Day Lanier. *Acrylic, charcoal, pastel, glue, collage, spray paint, 41 1/2 × 31 1/8. 1977. [Jim Dine Figure Drawings. Constance W. Glenn. p. 73.]*

Figure 194 *Hedda Stern.* Richard Linder. *72″ × 60.″ 1967. Parsons Gallery.*

Figure 195 *Paul Delaroche (1869–1948)* Oliver Cromwell. *Victoria & Albert Museum. GF 1548.*

*Figure 196*    *Leonardo da Vinci (1452–1519).* Profile of a Man.
*10.2 × 7.4 cm. plate CV, Cat. #111. British Museum. [14th/15th*
Century Italian Drawings Catalog. *1950.]*

*Figure 197*    *Lucian Freud.* Frank Auerbach. *Oil on canvas.*
*1975–76. [Art in America,* May 1988. *p. 136. Text on p. 133.]*

**Figure 198** *Stanley Spencer R. A.* Self portrait. *Oil on canvas, 24 3/4 × 20. From Trustees of Tate Gallery. 1914. [Stanley Spencer. Royal Academy of Art. p. 50.]*

**Figure 199** *Michelangelo (1475–1564).* Portrait of Andrea Quaratesi. *20.5 × 16.5 cm. From the Ashmolean Museum (Parker 315). [Drawings by Michelangelo. British Museum. p. 93.]*

**Figure 200** *John Hamil for Martiner (1741–1779).* Head. *#855. The Courtauld: London, England.*

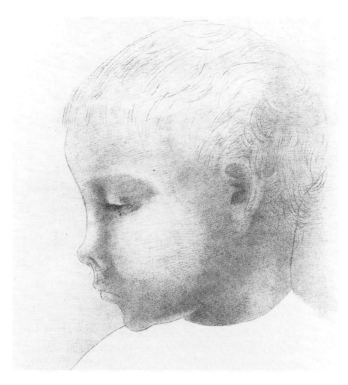

**Figure 201**    *Odilon Redon. (1840–1916).* Mon Enfant (Portrait of Ari). *1892 #124. Lithograph, 230 × 217 cm. National Gallery of Art, Washington, Ailsa Mellon Bruce Fund.*

**Figure 202**    *Sebastiano del Piombo.* Head of a Man Looking Up. *38 × 24.7 cm. [British Museum, London, England, 14/15 century Italians.]*

**Figure 203**    *Giovanni Bellini. (Attributed to).* Man's Head. *38.9 × 26.1 cm. [Plate XV. Catalog #16. British Museum, London, England. 14/15 Century Italians.]*

171

*Figure 204*    *Permeke.* Nude. *Oil on paper, 57 1/2" × 37 1/2". From M. Naessans, Miese. 1921. [*Ensor to Permeke, Royal Academy of Arts, *p. 68 & 41.]*

*Figure 205*    *Otto Dix (1891–1969).* Standing Nude. *Pencil, 28 3/8" × 19 1/4". From Collection of Alfred Hrdlicka, Vienna, Austria. [*Naked to Nude, *by Georg Eisler.]*

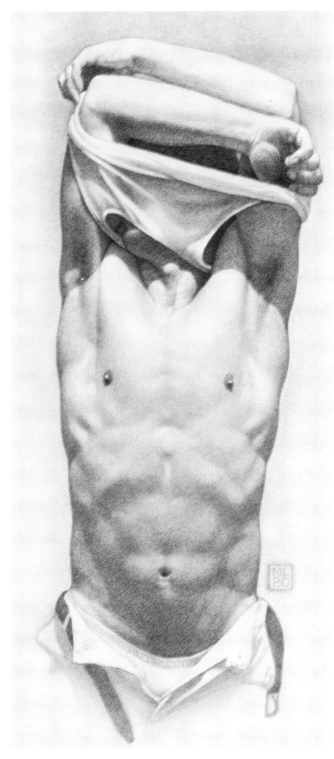

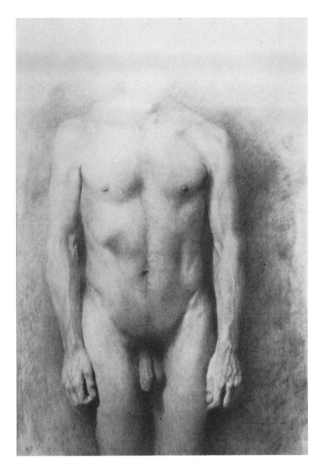

*Figure 207*   *Francisco Lopez, (1932– ).* Male Nude. *Pencil, 42 1/2'' × 30 3/4'' (108 × 78 cm.). From Galerie Meyer-Ellinger, Frankfurt am Main. 1973. [*Naked to Nude, *by George Eisler.]*

*Figure 206*   *Michael Leonard, (1933– ).* Stripped Torso 6. *Pencil 11 1/4'' × 4 3/8''. From private collection. 1980. [*The Body *by Edward Lucie Smith, Thames & Hudson, 500 5th Ave, N.Y.]*

173

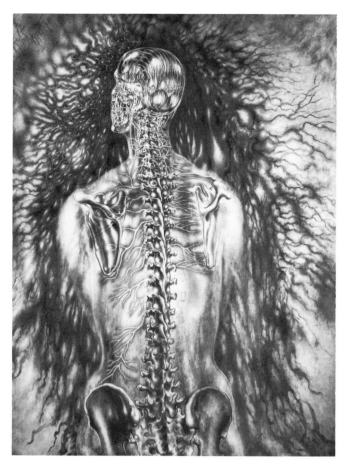

**Figure 208** *Pavel Tchelitchew. Anatomical Painting. © 1944–45 Oil on canvas, 56 × 46" (142.2 × 116.8 cm) Collection of Whitney Museum of American Art, New York. Gift of Lincoln Kirstein Accession number 62.26.*

**Figure 209** Michelangelo (1475–1564). Studies for Libyan Sibyl. *Red chalk, 29 × 21.5 cm. From The Metropolitan Museum of Art, New York. [Lit: Dussler 339; Hartt 87. Drawings by Michelangelo.* British Museum. *p. 31.]*

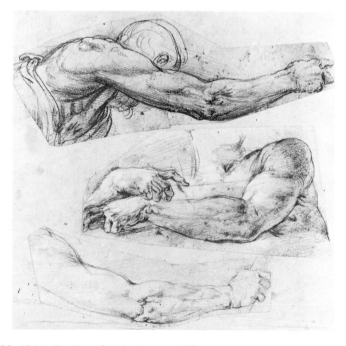

**Figure 210** Battista Franco (1498–1561). Studies of Male Arms and Thorax. *Red chalk (on one sheet inked over), 7" × 7 1/2" (177 × 190 cm.). Victoria and Albert Museum [Catalog Victoria and Albert, vol I & II. By Peter Ward. #137.]*

*Figure 211*   *Michelangelo (1475–1564).* Study of a Male
Nude. *Black chalk, 29.3 × 23.3 cm.* [Drawings by Michelangelo,
*British Museum, p. 121. #137. Lit: Dussler 563; Tolnay, v, p. 190, no.
185; Hartt 390.*]

*Figure 212*   Study for the Figure of Christ on the Cross.
*Black chalk heightened with white and some brown wash along the
outline of the right arm. 52.8 × 37 cm. Provenance: R. Payne
Knight bequest, 1824 Oo. 9–26 Department of Prints and
Drawings, British Museum, London, England. [Artist's Source:*
Reubens Drawings and Sketches, *British Museum, p. 50.*]

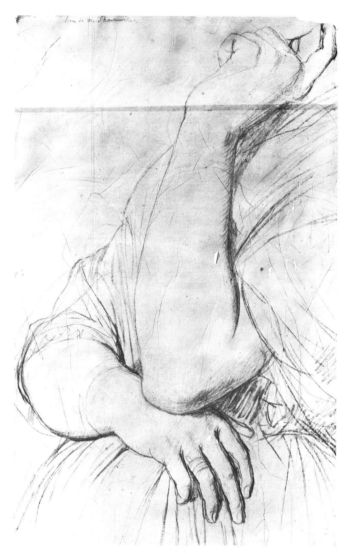

**Figure 213** *Jean-Auguste Ingres (1780–1867).* The Arms of
the Contesse d'Haussonville. *#35. Black chalk, 470 × 309 cm.
867.262; M. 1829; Ternon's, No. 69; Paris No. 236. [Ingres, Arts Council
of Great Britian. p. 31.]*

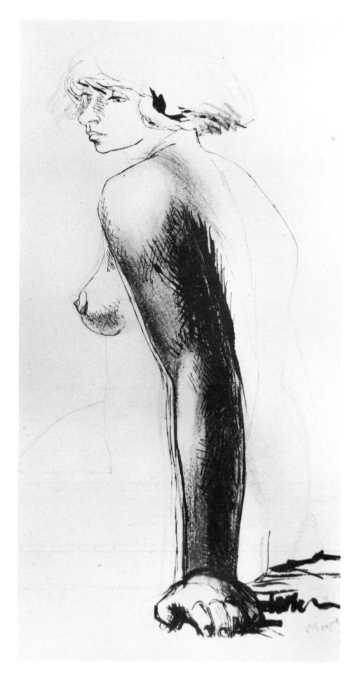

**Figure 214** *Henry Moore (1898– ).* Seated Nude. *Ink and
crayon, 41.9 × 33.8 cm. From Mr. and Mrs. Gordon Bunshaft. 1927.
[*Naked to Nude, *by Georg Eisler.]*

**Figure 215** *Edgar Degas (1834–1917).* Four Studies of a
Dancer. *19 3/4 × 12 1/2. From Musee de Louvre, Paris, France.*
*[Arts Council Catalog, #1, G.13.]*

**Figure 216** Studies of Hands. *Oxford Ashmolean Museum.*
*Dept. of Western Art. E D B 156 LELY*

Figure 217    Daniel Quintero. Dead Hands. *Pencil, 7 1/4" × 9 5/8" (18.5 × 24.5 cm). 1974. [Fine Art Library Catalog (#16—?) London.]*

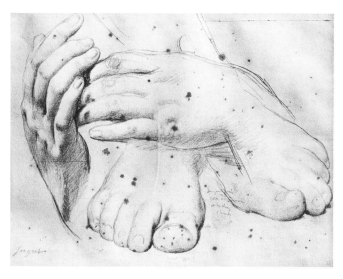

Figure 219    *Jean-Auguste Ingres (1780–1867).* Study for a woman's two hand and two feet, No. 24. *Black chalk on blue paper, 9 3/4" × 13 5/8" (249 × 345 cm). [Arts Council, Great Brit. #1, Ingres Catalogue, 1957, p. 18. Madame Ingres bequest, 1886. Studies for definitive version of the "Verso of Louis VIII."]*

Figure 218    *Andrea del Sarto.* Studies of Hands. *Chalk. From The Uffizi, Florence, Italy. [Anatomy Lessons, Coyle/Hale. p. 179.]*

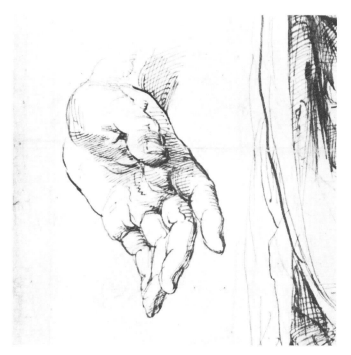

Figure 220    *Raphael Sanzio.* Drapery of Horace and Other Studies. (Detail) *Pen, brown ink over black chalk, 13 1/2" × 9 1/2". From British Museum, London, England [Anatomy Lessons, Coyle/Hale p. 173.]*

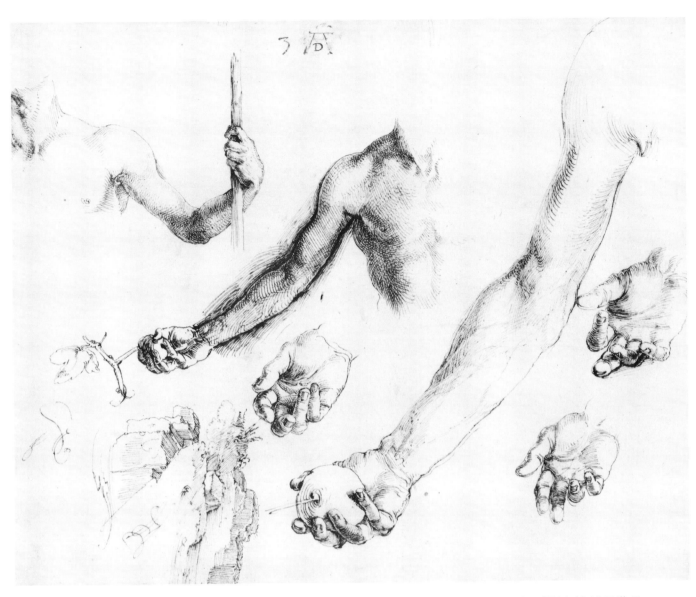

*Figure 221*    *Albrecht Durer (1471–1528).* Three studies from nature for Adam's Hand Engraving, *8 1/2″ × 10 13/16″. From British Museum, London, England. [Anatomy Lessons,* Coyle/Hayle, p. 155.*]*

**Figure 222** *Raphael (1483–1520). Study for 129. Lille, Musée Wicar. Pen and ink. 31.5 × 22.2. F.V. 245, m. 54. Devonshire Collection, Chatsworth, Reproduced by permission of the Chatsworth Settlement Trustees.*

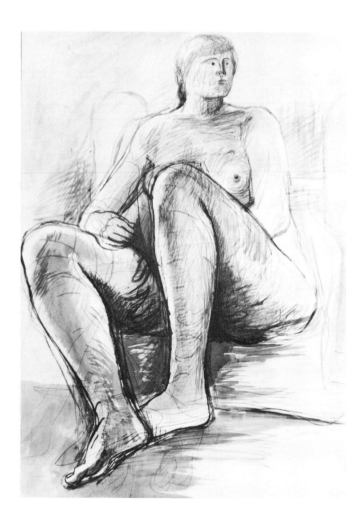

**Figure 223** *Henry Moore (1898– ). Drawing for Life. Seated figure. Ink, wash, and pencil. 22″ × 15″ (55.9 × 35.1 cm). From Collection of Dr. Alan Wilkinson, Toronto, Canada. [*Naked to Nude, *by Georg Eisler.]*

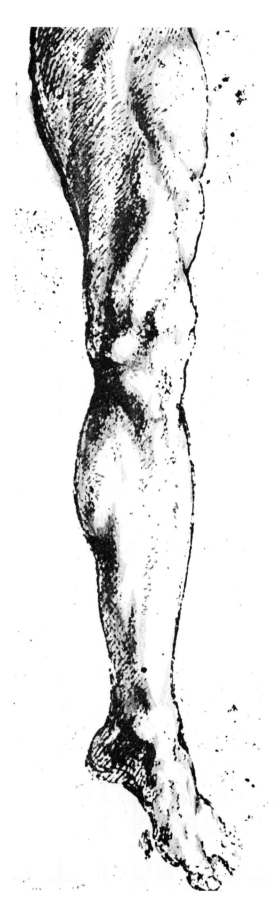

**Figure 224** *Michelangelo (1475–1564). Study of the legs of a* nude man. *Black chalk. Library of Christ Church, Oxford, England.* [Anatomy Lessons, *Coyle/Hale, p. 59.]*

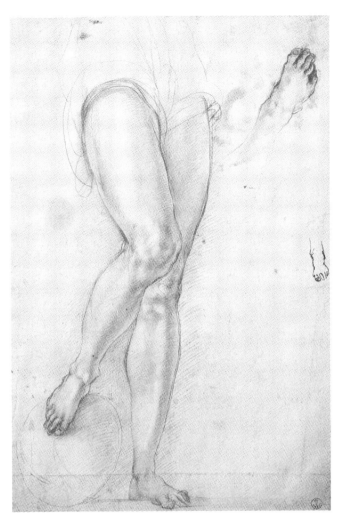

**Figure 226**  *Anonymous Copy.* Drawn from the Medici Chapel. *Red chalk, 34 × 26.2 cm. Plato CLII Cat. G021.*

**Figure 225**  *Jacopo Pontormo (1494–1556).* Study of the lower part of female nude. *Red chalk, 15 9/16″ × 10 1/4″. From The Uffizi, Florence, Italy. [Anatomy Lessons, Coyle/Hale, p. 91.] [Michelangelo and Sundry, British Museum, Italian Drawings, 1953. #20.]*

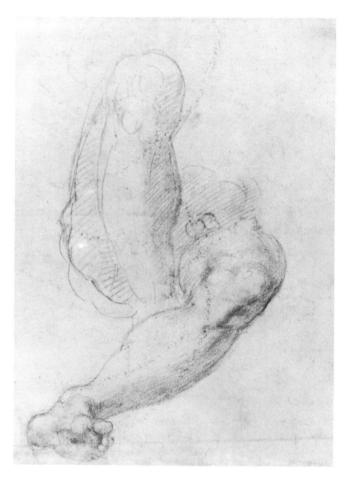

**Figure 227** *Michelangelo (1475–1564). Verso. [Catalogue of Collection of Drawings of Ashmolean Museum, Vol. II. Plate #330.]*

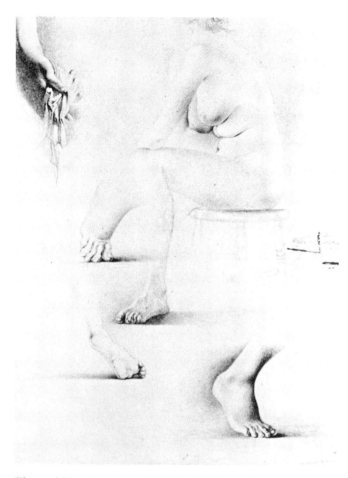

**Figure 228** *Salvador Dali (1904–1989). Studies of a Nude. Pencil. From John Thrall Soby Collection Museum of Modern Art, New York. 1935. [Seurat to Matisse: Drawing in France, by William S. Lieberman, p. 70.]*

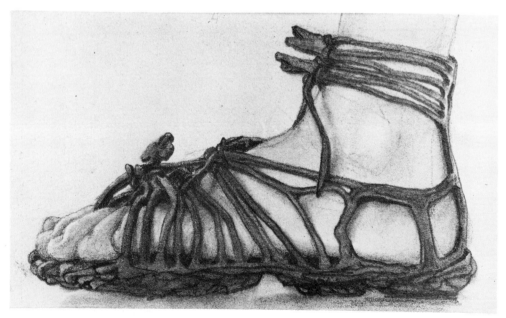

**Figure 229** *Sir Lawrence Alma-Tadema (1836–1912).* Sandalled Foot. *450 42 Victoria & Albert Museum, London, England. E 2606- 1915.*

**Figure 231** *Peter Paul Reubens (1577–1640).* Studies for three figures (in the miracle of St. Xavier). *Detail of feet and legs. W. S. II. Victoria and Albert Museum, London, England. GD 3636 D904, 905–1900.*

**Figure 230** *Jack Stuck. Unknown title. [*Art Forum, *Feb. 1988.]*

**Figure 232** *Rene Magritte (1898–1967).* The Red Model. *Oil on canvas, 21 15/16″ × 18 1/4″. From Collection of the Musee National d'Art Moderne, Paris. 1935. [*Art Forum, *Mar 1988.]*

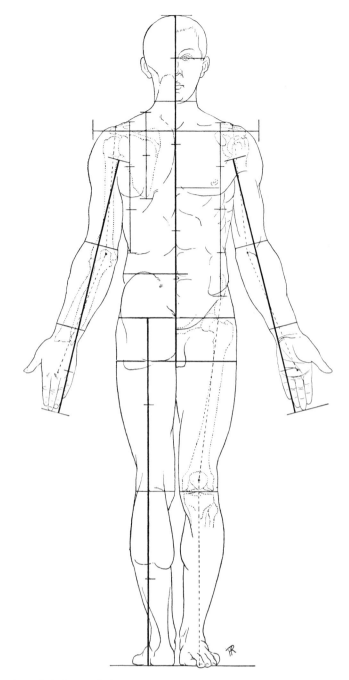

**Figure 233**   Paul Richer. *Proportions of the human body. [Artistic Anatomy. Translated and edited by Robert Beverly Hale. Watson-Guptill Publ., NY. 1986. p. 248.]*

# PEOPLE AND PARTS

▼

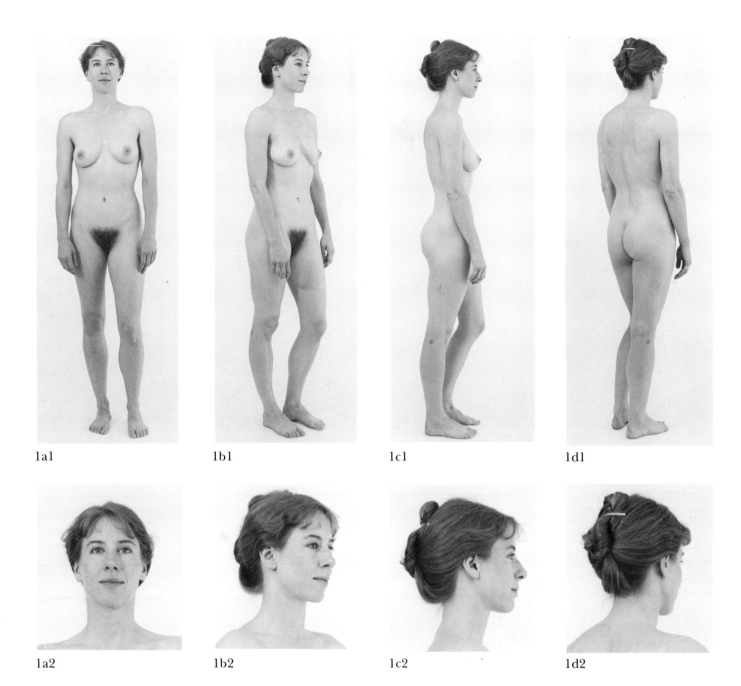

1a1

1b1

1c1

1d1

1a2

1b2

1c2

1d2

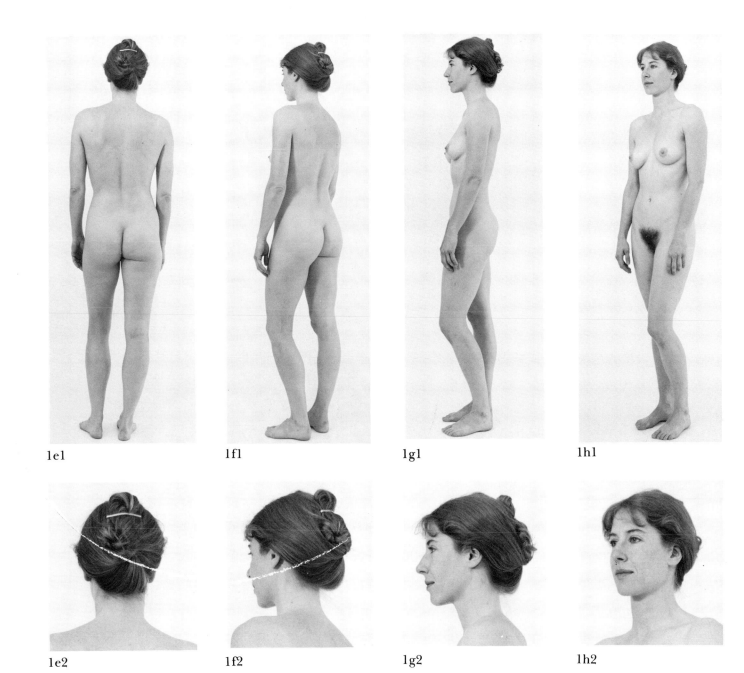

1e1

1f1

1g1

1h1

1e2

1f2

1g2

1h2

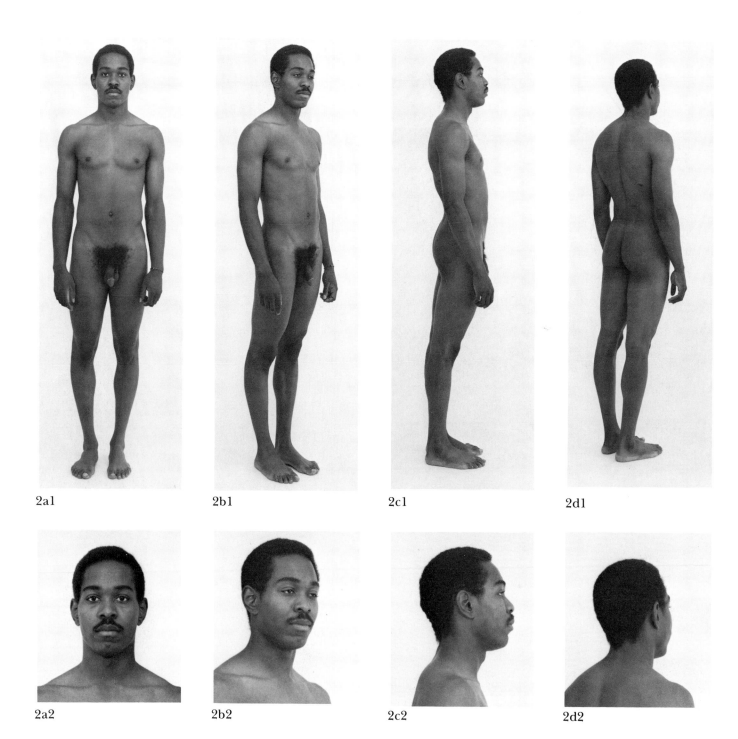

2a1

2b1

2c1

2d1

2a2

2b2

2c2

2d2

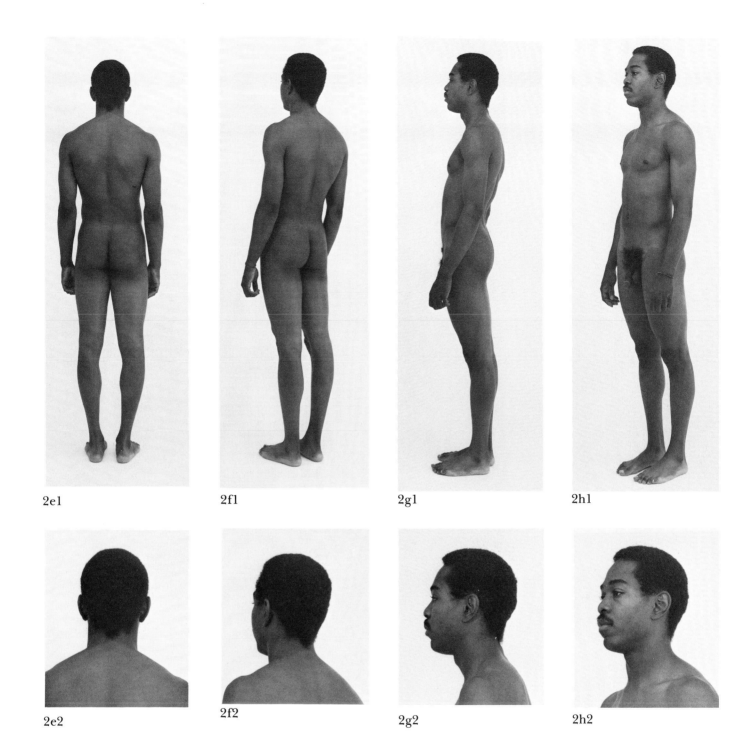

2e1

2f1

2g1

2h1

2e2

2f2

2g2

2h2

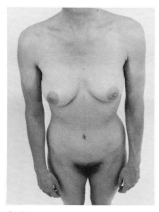

3a1

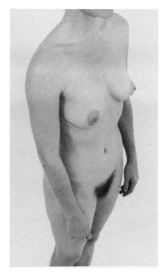

3b1

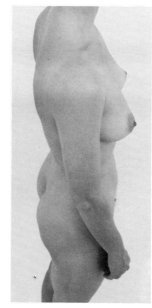

3c1

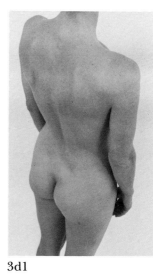

3d1

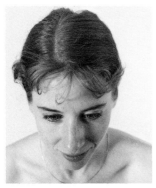

3a2

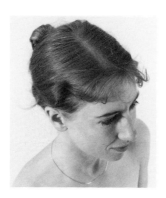

3b2

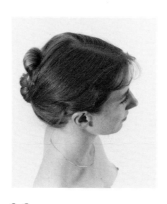

3c2

3d2

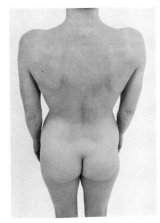

3e1

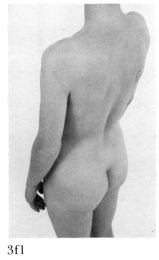

3f1

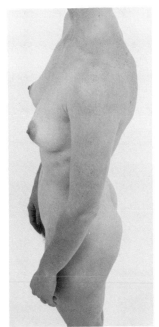

3g1

3h1

3e2

3f2

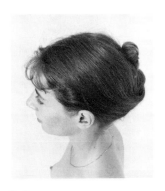

3g2

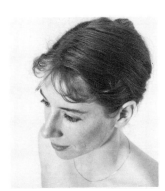

3h2

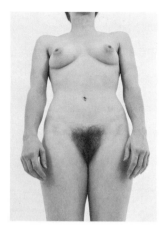

4a1

4b1

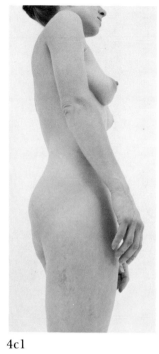

4c1

4d1

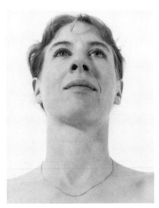

4a2

4b2

4c2

4d2

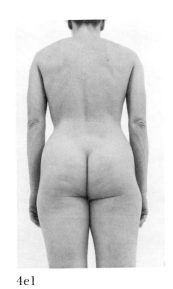

4e1

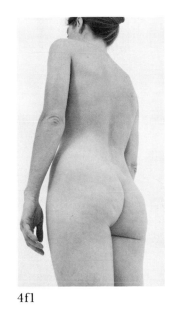

4f1

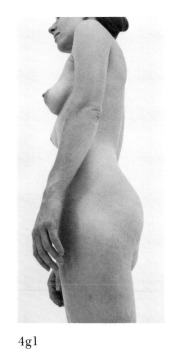

4g1

4h1

4e2

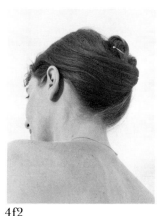

4f2

4g2

4h2

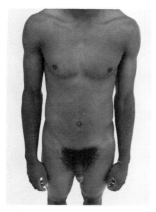

5a1

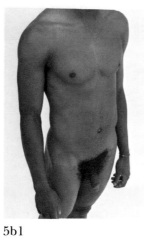

5b1

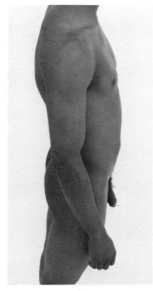

5c1

5d1

5a2

5b2

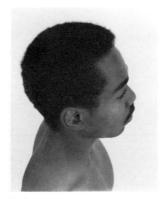

5c2

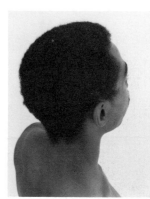

5d2

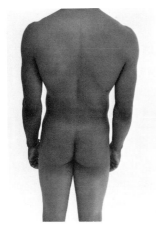

5e1

5f1

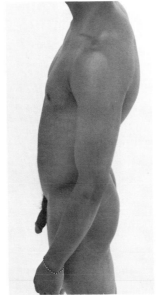

5g1

5h1

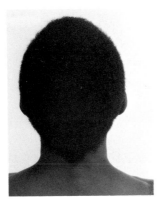

5e2

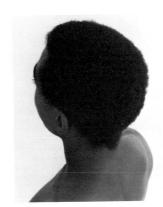

5f2

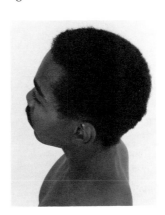

5g2

5h2

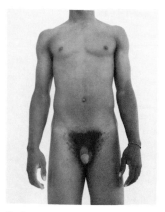

6a1

6b1

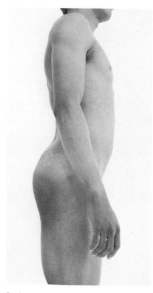

6c1

6d1

6a2

6b2

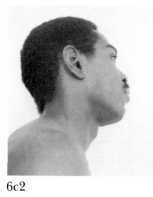

6c2

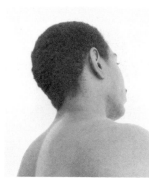

6d2

6e1

6f1

6g1

6h1

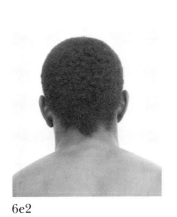

6e2

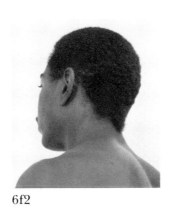

6f2

6g2

6h2

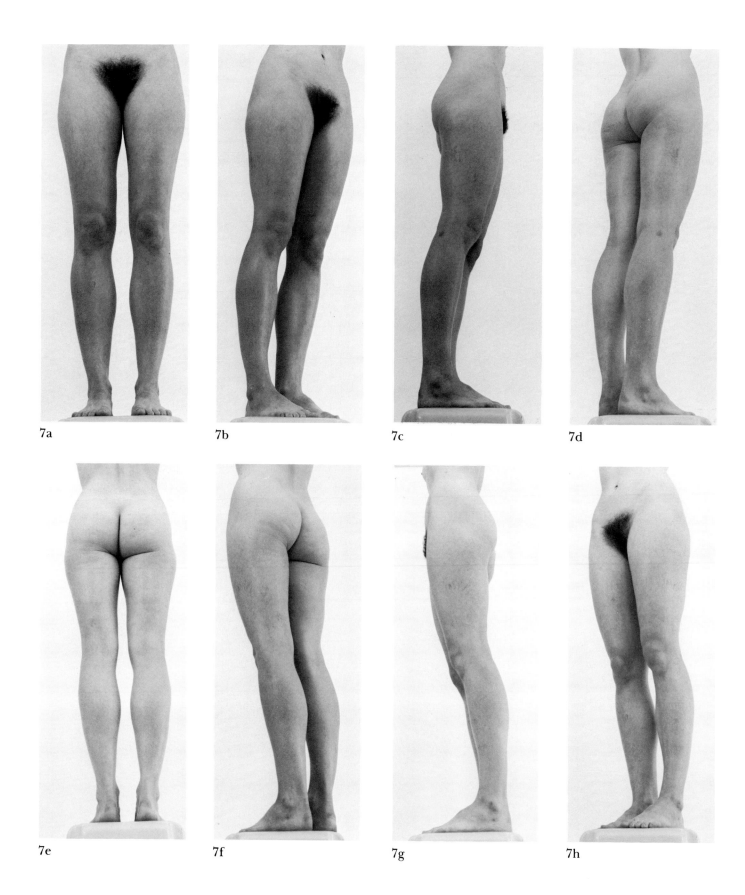

7a

7b

7c

7d

7e

7f

7g

7h

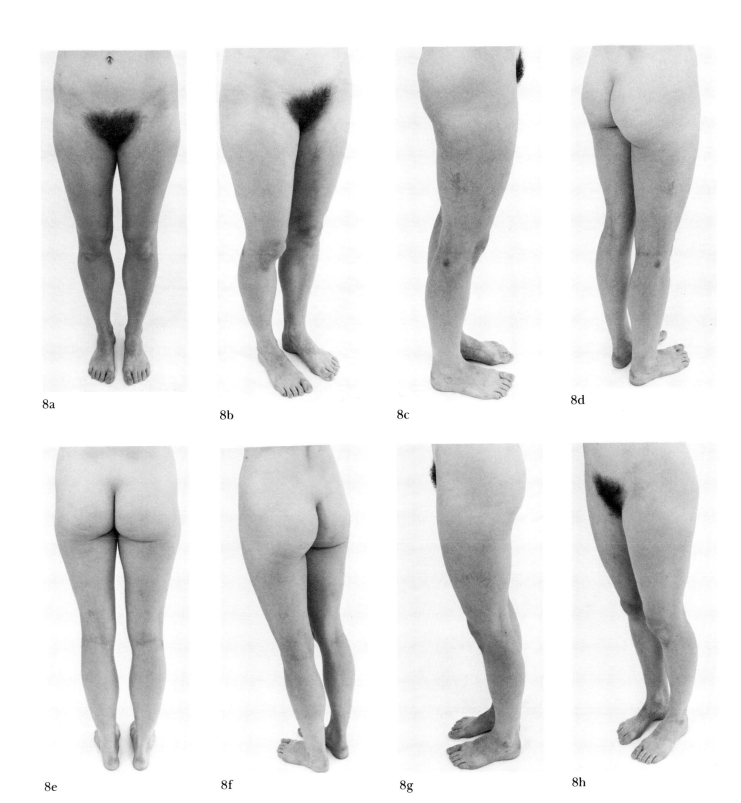

8a

8b

8c

8d

8e

8f

8g

8h

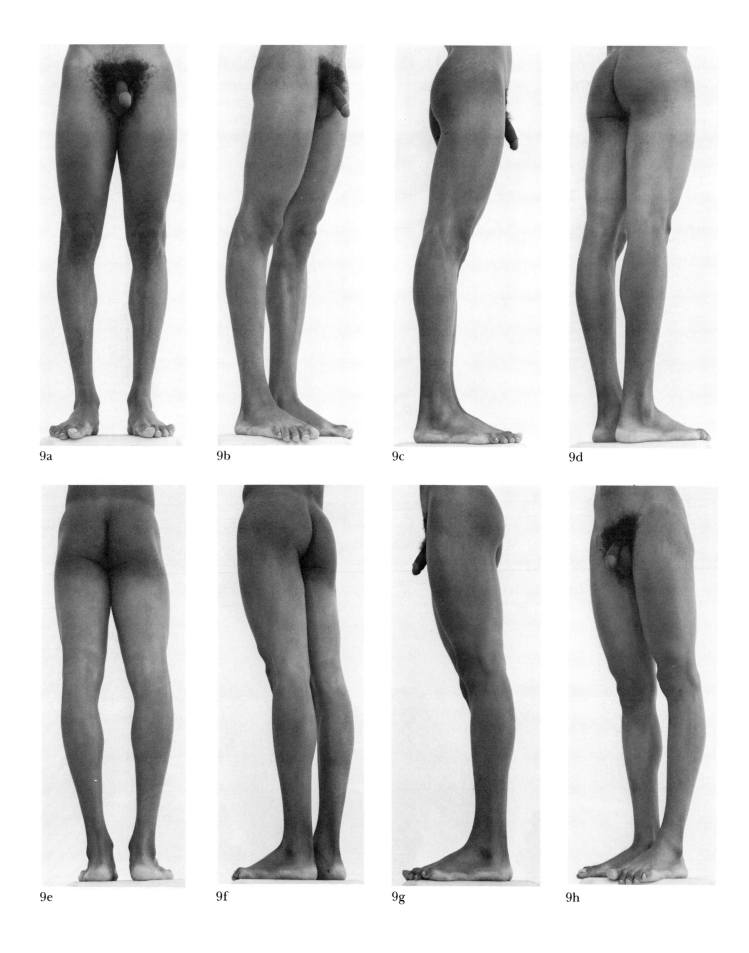

9a

9b

9c

9d

9e

9f

9g

9h

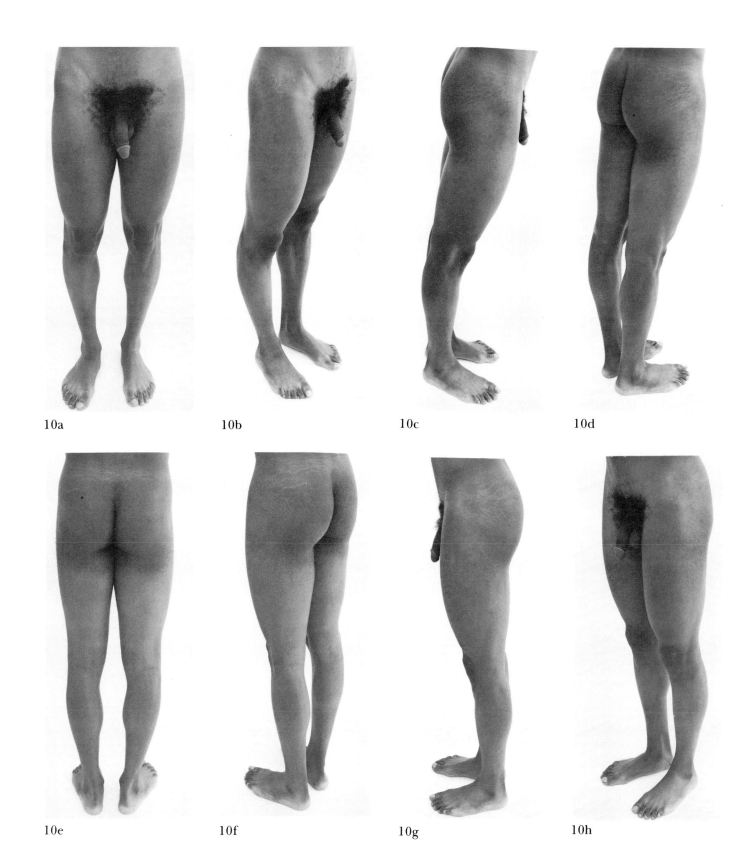

10a

10b

10c

10d

10e

10f

10g

10h

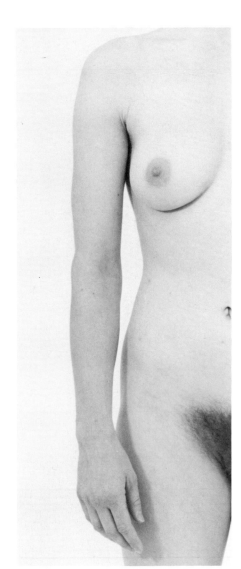
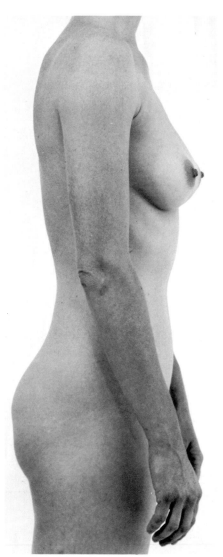
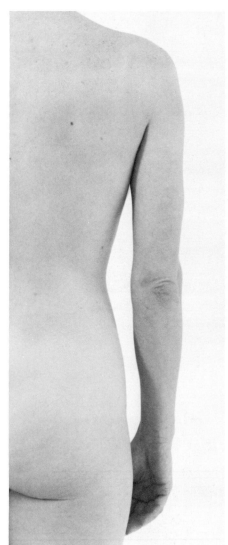

11a                              11b                              11c

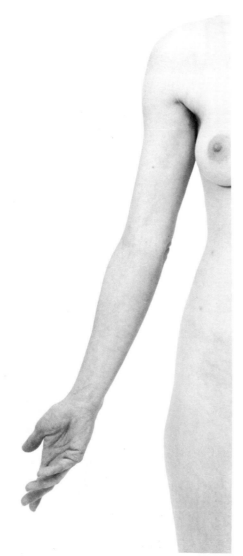
12a

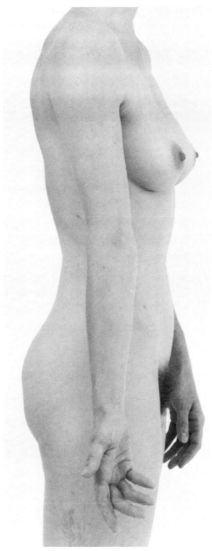
12b

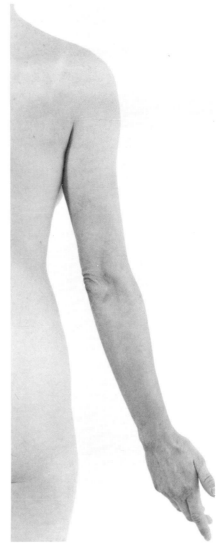
12c

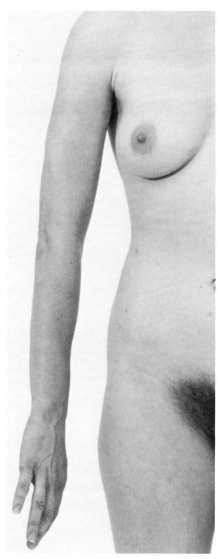
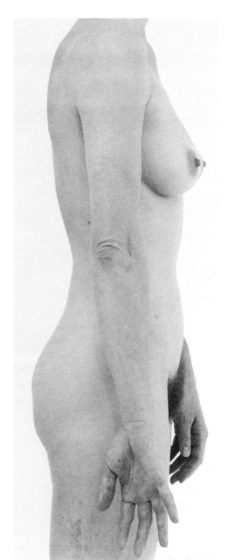
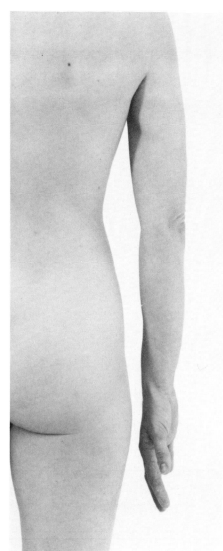

13a        13b        13c

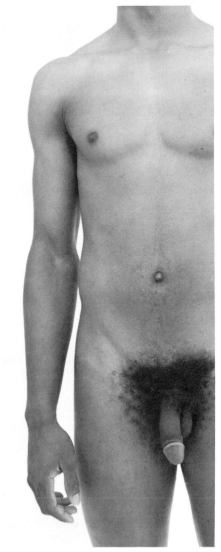

14a

14b

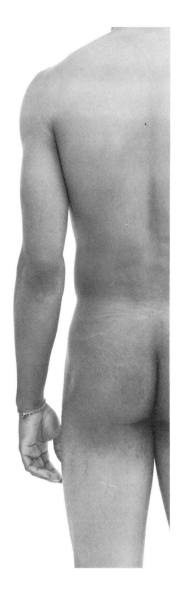

14c

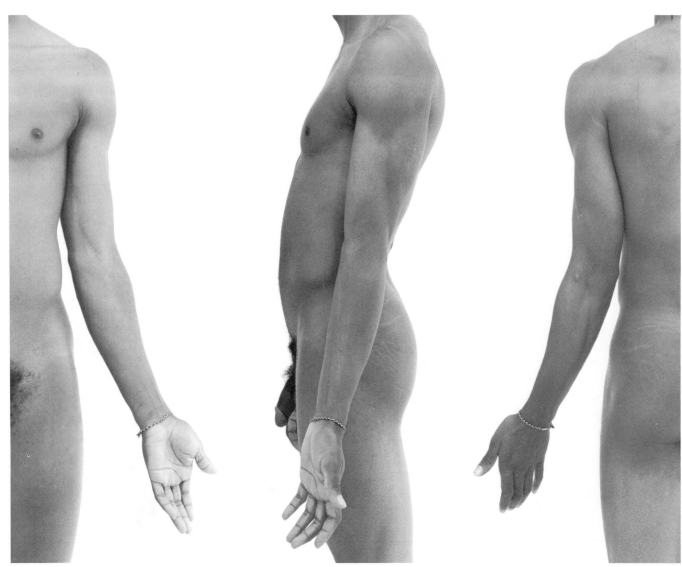

15a

15b

15c

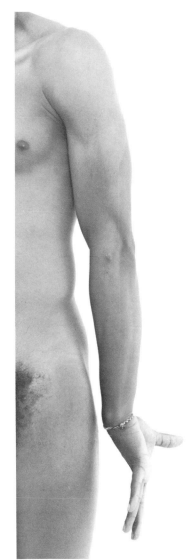

16a

16b

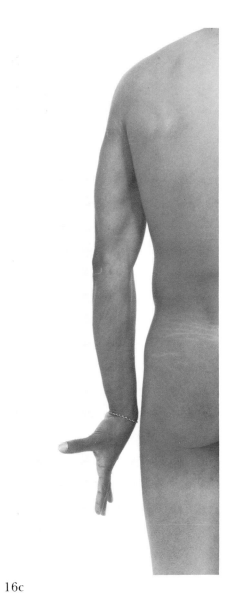

16c

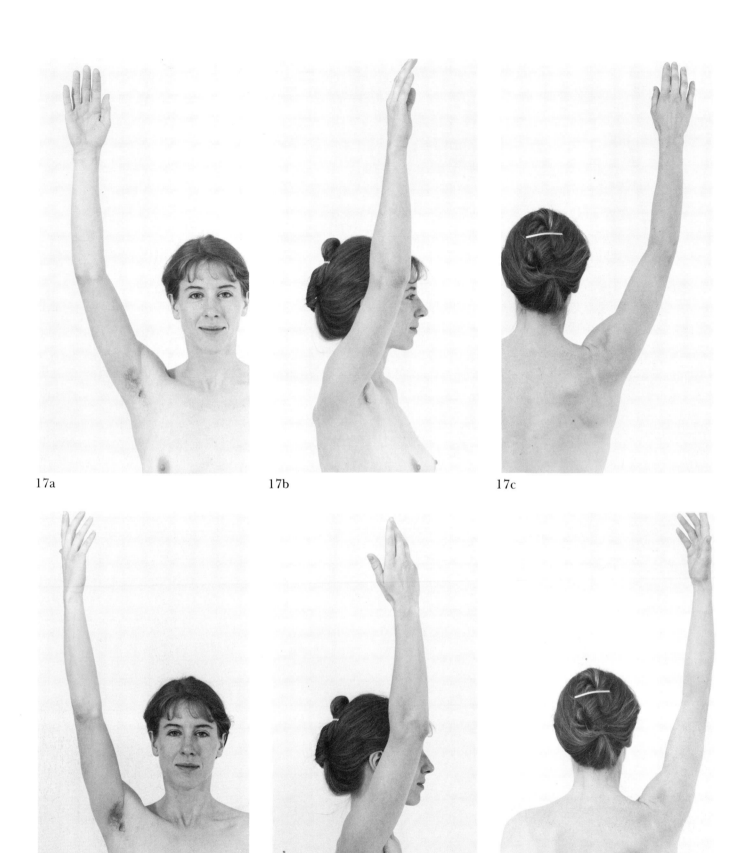

17a

17b

17c

18a

18b

18c

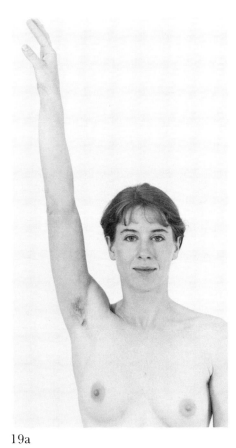

19a

19b

19c

20a

20b

20c

21a

21b

21c

22a

22b

22c

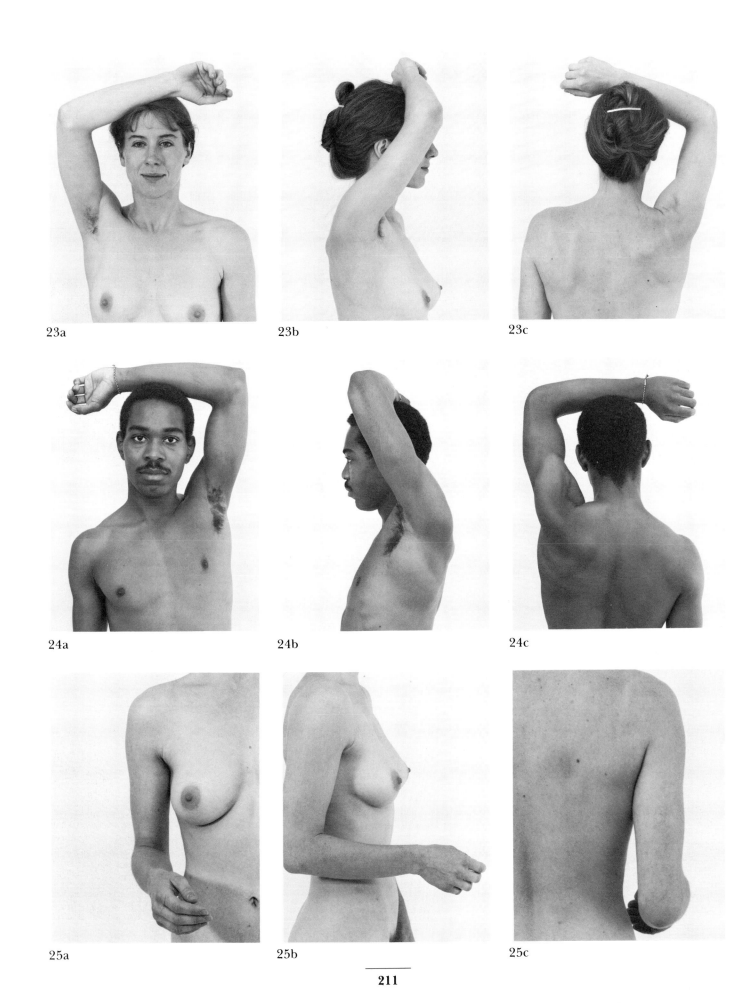

23a

23b

23c

24a

24b

24c

25a

25b

25c

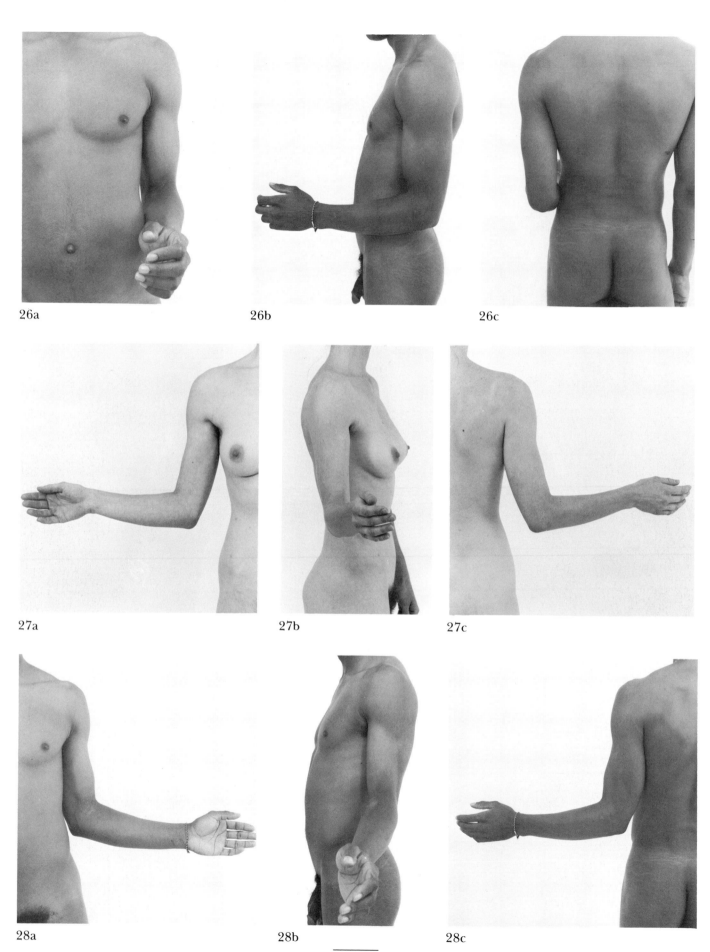

26a

26b

26c

27a

27b

27c

28a

28b

28c

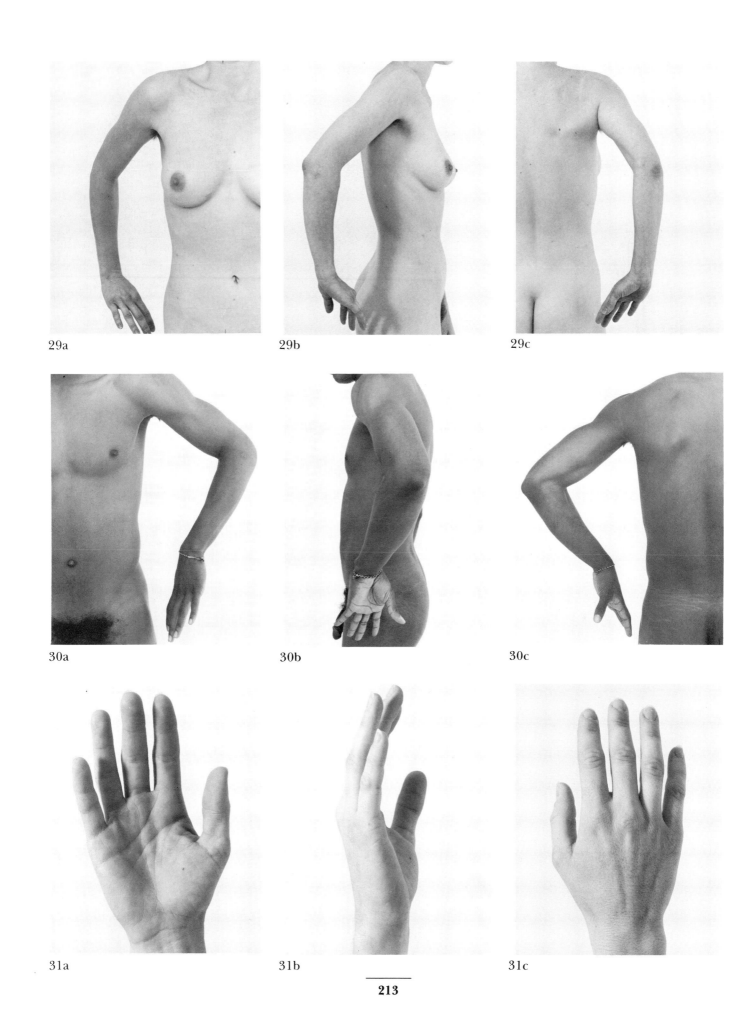

29a

29b

29c

30a

30b

30c

31a

31b

31c

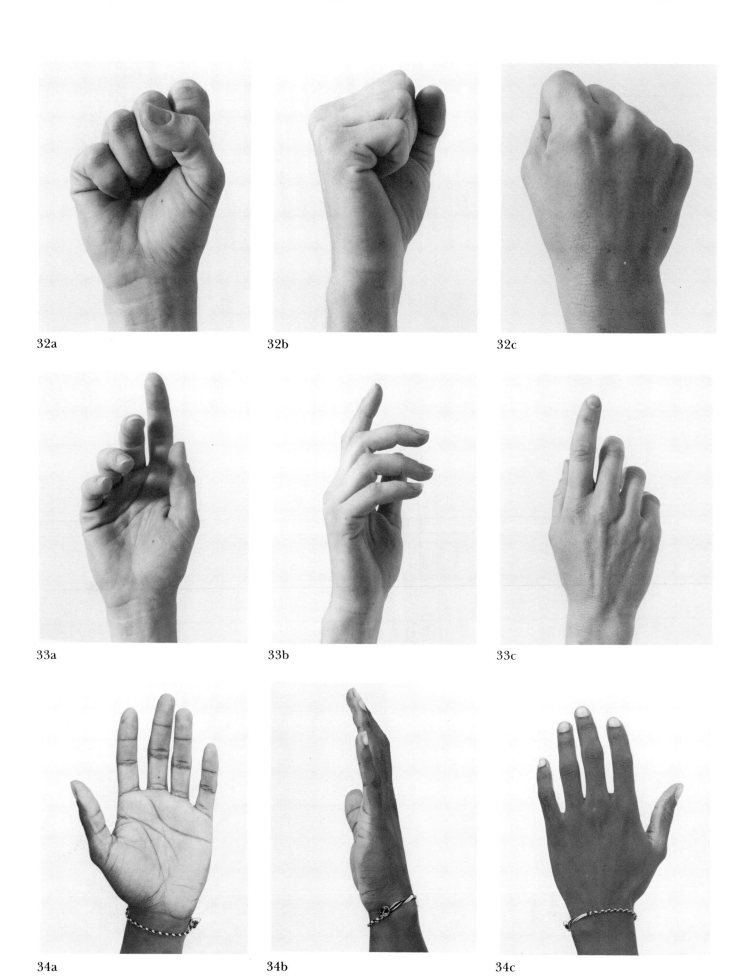

32a            32b            32c

33a            33b            33c

34a            34b            34c

35a

35b

35c

36a

36b

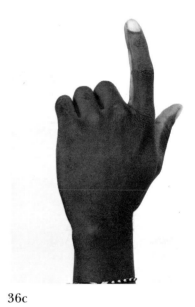

36c

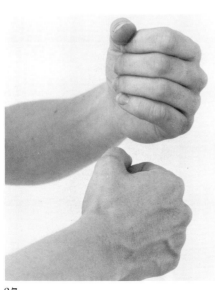

37a

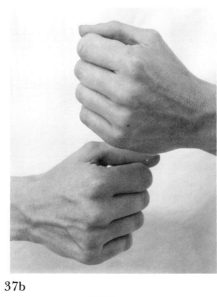

37b

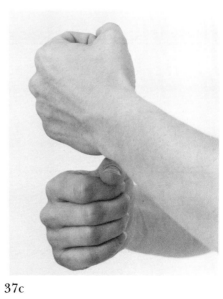

37c

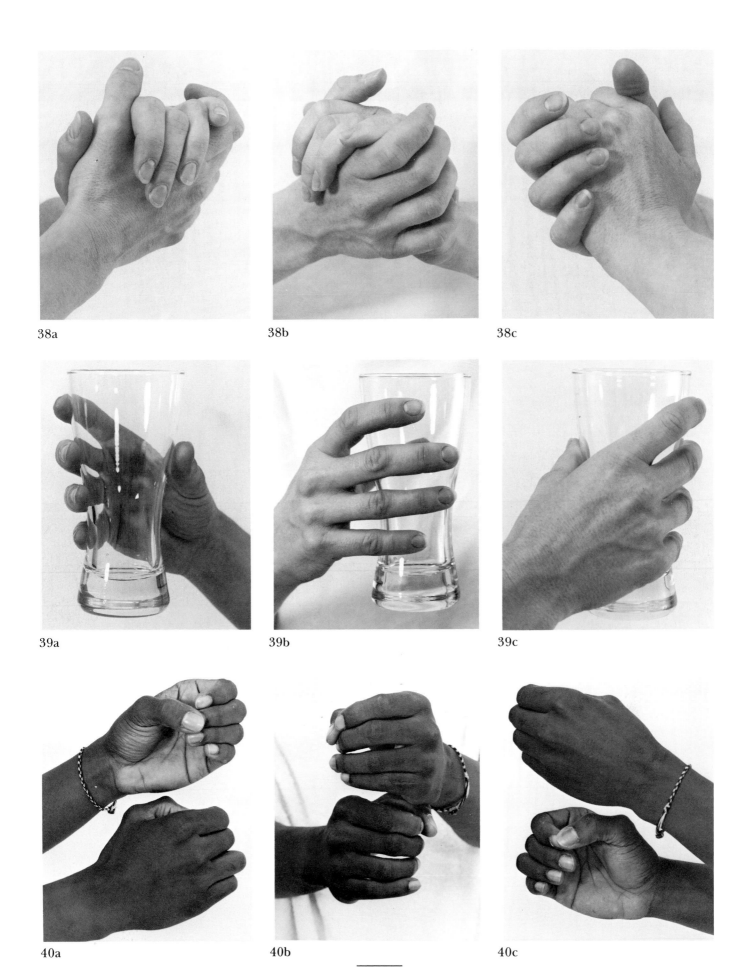

38a

38b

38c

39a

39b

39c

40a

40b

40c

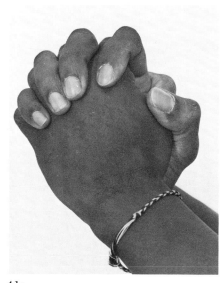

41a

41b

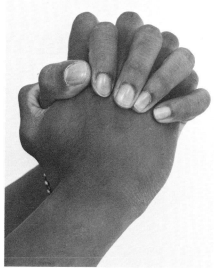

41c

42a

42b

42c

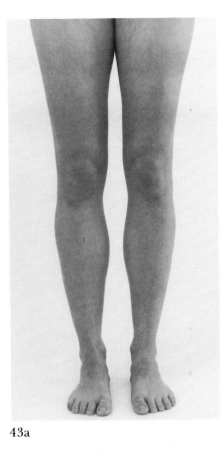

43a

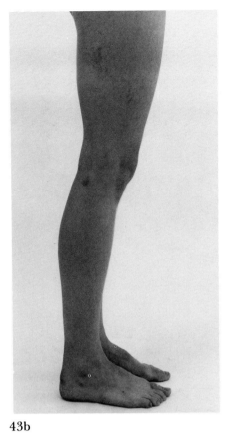

43b

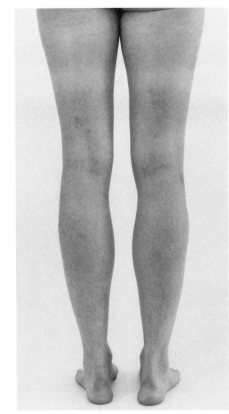

43c

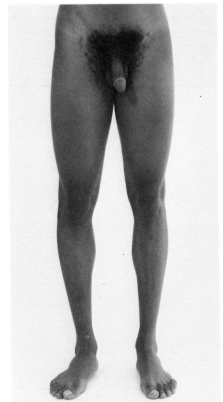

44a

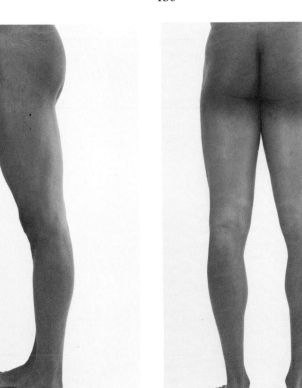

44b

44c

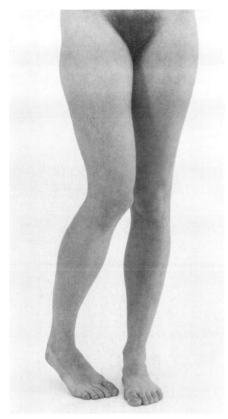

45a

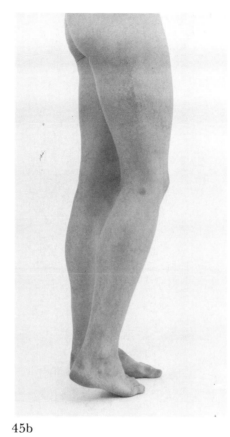

45b

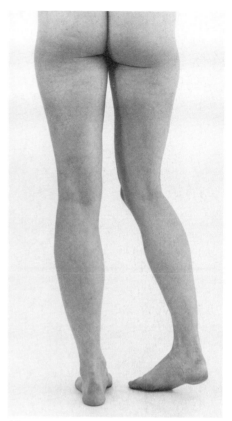

45c

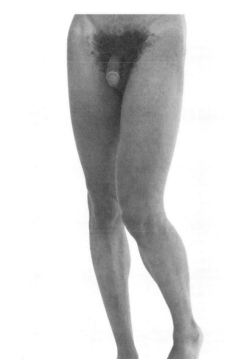

46a

46b

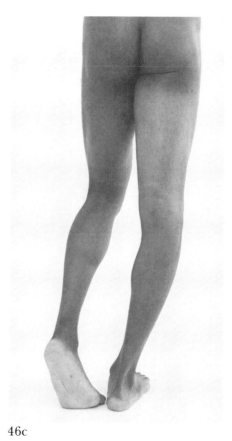

46c

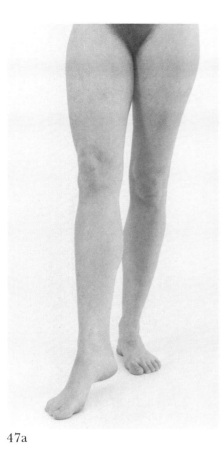

47a

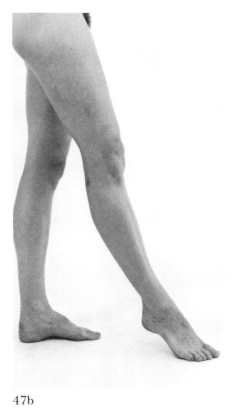

47b

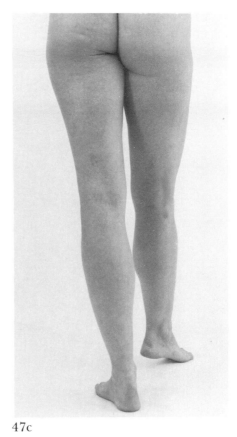

47c

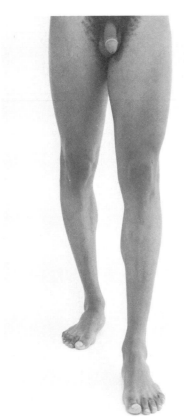

48a

48b

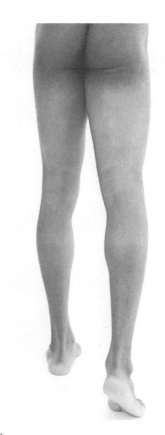

48c

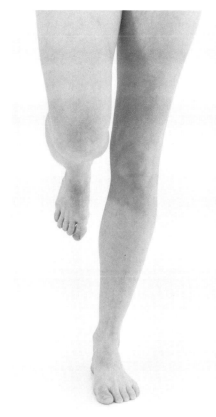

49a

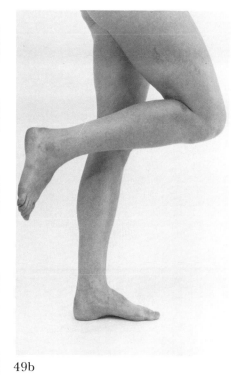

49b

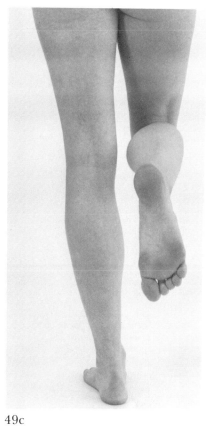

49c

50a

50b

50c

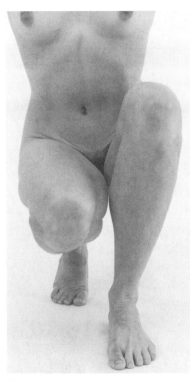

51a

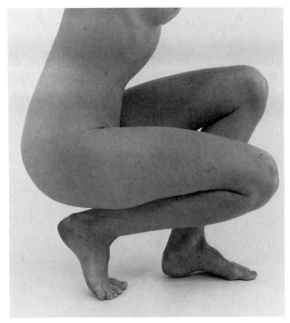

51b

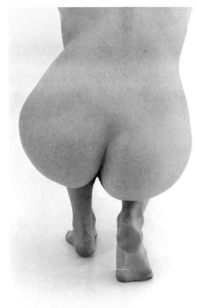

51c

52a

52b

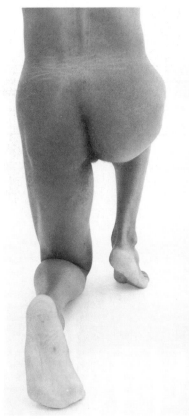

52c

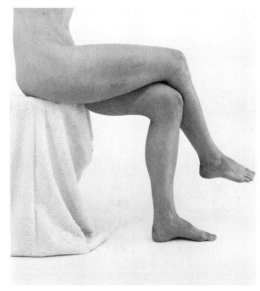

53a

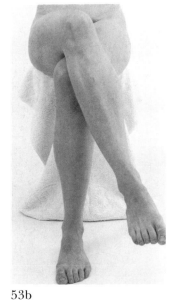

53b

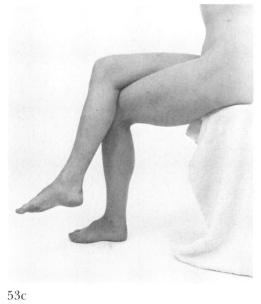

53c

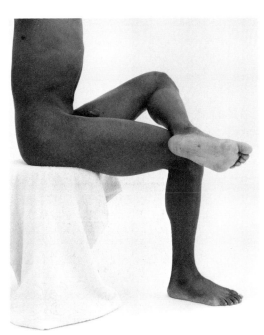

54a

54b

54c

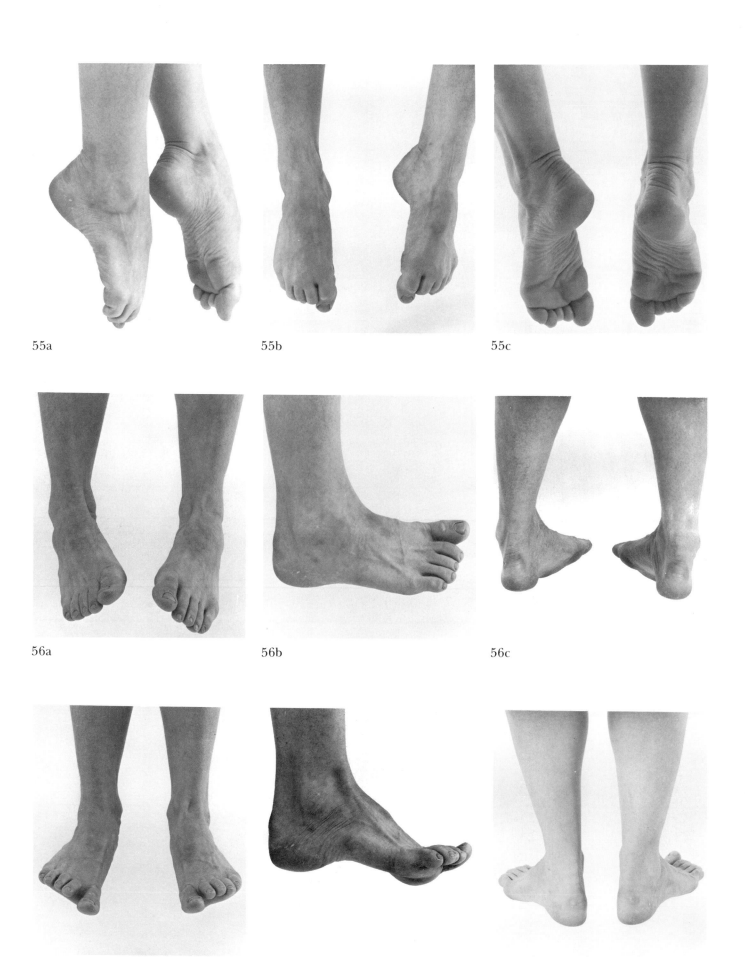

55a

55b

55c

56a

56b

56c

57a

57b

57c

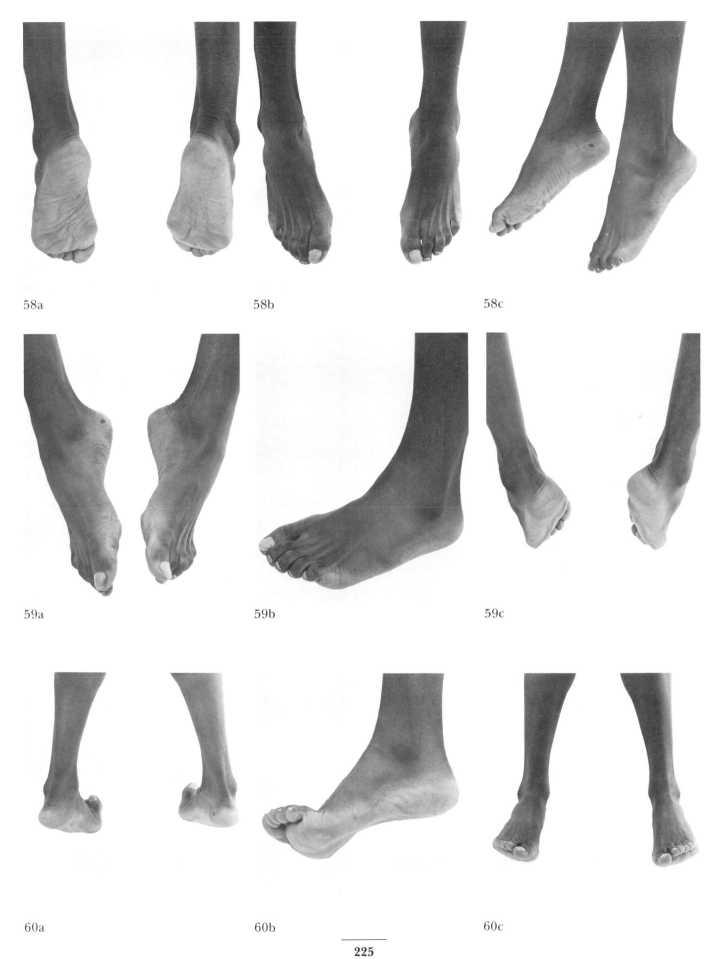

58a

58b

58c

59a

59b

59c

60a

60b

60c

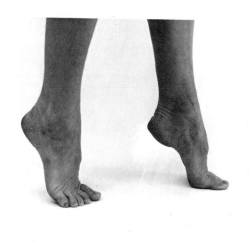

61a

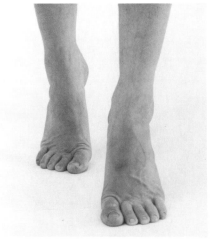

61b

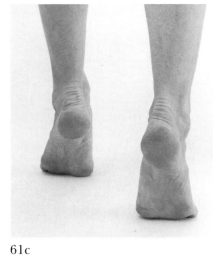

61c

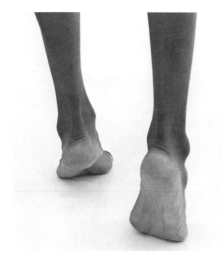

62a

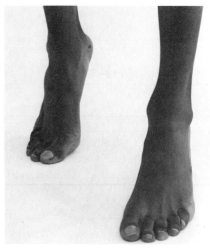

62b

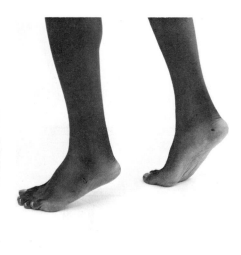

62c